Pyrosequencing® Protocols

METHODS IN MOLECULAR BIOLOGY™

John M. Walker, SERIES EDITOR

387. Serial Analysis of Gene Expression (SAGE): *Digital Gene Expression Profiling,* edited by *Kare Lehmann Nielsen, 2007*

386. Peptide Characterization and Application Protocols, edited by *Gregg B. Fields, 2007*

385. Microchip-Based Assay Systems: *Methods and Applications,* edited by *Pierre N. Floriano, 2007*

384. Capillary Electrophoresis: *Methods and Protocols,* edited by *Philippe Schmitt-Kopplin, 2007*

383. Cancer Genomics and Proteomics: *Methods and Protocols,* edited by *Paul B. Fisher, 2007*

382. Microarrays, Second Edition: *Volume 2, Applications and Data Analysis,* edited by *Jang B. Rampal, 2007*

381. Microarrays, Second Edition: *Volume 1, Synthesis Methods,* edited by *Jang B. Rampal, 2007*

380. Immunological Tolerance: *Methods and Protocols,* edited by *Paul J. Fairchild, 2007*

379. Glycovirology Protocols, edited by *Richard J. Sugrue, 2007*

378. Monoclonal Antibodies: *Methods and Protocols,* edited by *Maher Albitar, 2007*

377. Microarray Data Analysis: *Methods and Applications,* edited by *Michael J. Korenberg, 2007*

376. Linkage Disequilibrium and Association Mapping: *Analysis and Application,* edited by *Andrew R. Collins, 2007*

375. In Vitro Transcription and Translation Protocols: *Second Edition,* edited by *Guido Grandi, 2007*

374. Quantum Dots: *Applications in Biology,* edited by *Charles Z. Hotz and Marcel Bruchez, 2007*

373. Pyrosequencing® Protocols, edited by *Sharon Marsh, 2007*

372. Mitochondrial Genomics and Proteomics Protocols, edited by *Dario Leister and Johannes Herrmann, 2007*

371. Biological Aging: *Methods and Protocols,* edited by *Trygve O. Tollefsbol, 2007*

370. Adhesion Protein Protocols, *Second Edition,* edited by *Amanda S. Coutts, 2007*

369. Electron Microscopy*: Methods and Protocols, Second Edition,* edited by *John Kuo, 2007*

368. Cryopreservation and Freeze-Drying Protocols, *Second Edition,* edited by *John G. Day and Glyn Stacey, 2007*

367. Mass Spectrometry Data Analysis in Proteomics, edited by *Rune Matthiesen, 2007*

366. Cardiac Gene Expression: *Methods and Protocols,* edited by *Jun Zhang and Gregg Rokosh, 2007*

365. Protein Phosphatase Protocols: edited by *Greg Moorhead, 2007*

364. Macromolecular Crystallography Protocols: *Volume 2, Structure Determination,* edited by *Sylvie Doublié, 2007*

363. Macromolecular Crystallography Protocols: *Volume 1, Preparation and Crystallization of Macromolecules,* edited by *Sylvie Doublié, 2007*

362. Circadian Rhythms: *Methods and Protocols,* edited by *Ezio Rosato, 2007*

361. Target Discovery and Validation Reviews and Protocols: *Emerging Molecular Targets and Treatment Options, Volume 2,* edited by *Mouldy Sioud, 2007*

360. Target Discovery and Validation Reviews and Protocols: *Emerging Strategies for Targets and Biomarker Discovery, Volume 1,* edited by *Mouldy Sioud, 2007*

359. Quantitative Proteomics by Mass Spectrometry, edited by *Salvatore Sechi, 2007*

358. Metabolomics: *Methods and Protocols,* edited by *Wolfram Weckwerth, 2007*

357. Cardiovascular Proteomics: *Methods and Protocols,* edited by *Fernando Vivanco, 2006*

356. High Content Screening: *A Powerful Approach to Systems Cell Biology and Drug Discovery,* edited by *D. Lansing Taylor, Jeffrey Haskins, and Ken Guiliano, 2007*

355. Plant Proteomics: *Methods and Protocols,* edited by *Hervé Thiellement, Michel Zivy, Catherine Damerval, and Valerie Mechin, 2006*

354. Plant–Pathogen Interactions: *Methods and Protocols,* edited by *Pamela C. Ronald, 2006*

353. DNA Analysis by Nonradioactive Probes: *Methods and Protocols,* edited by *Elena Hilario and John. F. MacKay, 2006*

352. Protein Engineering Protocols, edited by *Kristian Müller and Katja Arndt, 2006*

351. *C. elegans:* *Methods and Applications,* edited by *Kevin Strange, 2006*

350. Protein Folding Protocols, edited by *Yawen Bai and Ruth Nussinov 2007*

349. YAC Protocols, *Second Edition,* edited by *Alasdair MacKenzie, 2006*

348. Nuclear Transfer Protocols: *Cell Reprogramming and Transgenesis,* edited by *Paul J. Verma and Alan Trounson, 2006*

347. Glycobiology Protocols, edited by *Inka Brockhausen-Schutzbach, 2006*

346. *Dictyostelium discoideum* Protocols, edited by *Ludwig Eichinger and Francisco Rivero, 2006*

345. Diagnostic Bacteriology Protocols, *Second Edition,* edited by *Louise O'Connor, 2006*

344. Agrobacterium Protocols, *Second Edition: Volume 2,* edited by *Kan Wang, 2006*

343. Agrobacterium Protocols, *Second Edition: Volume 1,* edited by *Kan Wang, 2006*

342. MicroRNA Protocols, edited by *Shao-Yao Ying, 2006*

341. Cell–Cell Interactions: *Methods and Protocols,* edited by *Sean P. Colgan, 2006*

340. Protein Design: *Methods and Applications,* edited by *Raphael Guerois and Manuela López de la Paz, 2006*

339. Microchip Capillary Electrophoresis: *Methods and Protocols,* edited by *Charles S. Henry, 2006*

METHODS IN MOLECULAR BIOLOGY™

Pyrosequencing® Protocols

Edited by

Sharon Marsh

Division of Oncology,
Washington University School of Medicine,
St. Louis, MO

Foreword by

Howard L. McLeod

UNC Institute for Pharmacogenomics
and Individualized Therapy,
University of North Carolina,
Chapel Hill, NC

HUMANA PRESS TOTOWA, NEW JERSEY

Softcover reprint of the hardcover 1st edition 2007
999 Riverview Drive, Suite 208
Totowa, New Jersey 07512

www.humanapress.com

This publication is printed on acid-free paper. ∞
ANSI Z39.48-1984 (American Standards Institute) Permanence of Paper for Printed Library Materials.

Cover artwork design: Maritha Lundin, Pyrosequencing AB - a subsidiary of Biotage AB.

Production Editor: Erika J. Wasenda

Cover design by Patricia F. Cleary

For additional copies, pricing for bulk purchases, and/or information about other Humana titles, contact Humana at the above address or at any of the following numbers: Tel.: 973-256-1699; Fax: 973-256-8341; E-mail: orders@humanapr.com; or visit our Website: www.humanapress.com

10 9 8 7 6 5 4 3 2 1

e-ISBN 1-59745-377-3

ISSN 1064-3745

Library of Congress Cataloging-in-Publication Data
Pyrosequencing protocols / edited by Sharon Marsh.
p. ; cm. -- (Methods in molecular biology ; 373)
Includes bibliographical references and index.
ISBN-13: 978-1-61737-693-1
e-ISBN-13: 978-1-59745-377-6
1. DNA--Analysis. 2. Nucleotide sequence. I. Marsh, Sharon.
II. Series: Methods in molecular biology (Clifton, N.J.) ; v. 373.
[DNLM: 1. Sequence Analysis, DNA--methods. W1 ME 9616J
v.373 2007 / QU 450 P998 2007]
QP624.P97 2007
611'.018166--dc22

2006014474

Foreword

Over the last 50 years, genetics has changed from an observational science to a discipline that can unveil mechanistic understanding of disease or therapeutic response. This revolution has been fueled by the discovery of the conformation of DNA, the development of polymerase chain reaction technology, and the tools and information generated by the Human Genome Project. This has been expanded to the areas of microbiology, botany, and veterinary medicine, where diagnosis, disease prediction, and therapy selection are among the many common practical applications.

A key tool for many areas of applied genomics has been Pyrosequencing®. This technology has distinguished itself from others through its robust ability to deliver complex microsequencing information from most any source of DNA (e.g., blood, leaves, effluent, formalin-fixed tissue, etc.). Pyrosequencing has also filled a useful niche in the low-to-moderate capacity area (e.g., 15 single nucleotide polymorphisms in 95 samples). This has become very important in an era where the development of genomic technologies is focused on whole genome approaches or similar broad strategies, rather than the more modest throughput needs of applied research. Since this represents the majority of investigations, the technology has become a mainstay of most genomics laboratories.

Pyrosequencing® Protocols provides a timely compilation of many of the critical uses of Pyrosequencing. The editor has assembled a "who's who" of Pyrosequencing expertise and the authors have delivered thoughtful, applied advice on how to optimize the use of this technology. I congratulate those involved in the construction of this text and welcome those about to gain from the insights availed therein.

Howard L. McLeod
Fred N. Eshelman Distinguished Professor
Director, UNC Institute for Pharmacogenomics and Individualized Therapy
University of North Carolina, Chapel Hill, NC

Preface

Since the release of the working draft of the human genome, there has been a need for improved technology and assay throughput to make use of the information now at our fingertips. *Pyrosequencing® Protocols* brings together world experts in fields where Pyrosequencing® has been an essential technology for achieving their goals.

The advantages of the Pyosequencing® system lie with the range of technological applications that can be applied to single nucleotide polymorphisms, insertion/deletions (indels), short tandem repeats, pooled allele frequencies, human leukocyte antigen typing, gene copy number, allelic imbalance in RNA, methylation status, and short sequencing stretches, many of which are described in this book. The assay is applicable to almost any source of DNA or RNA (e.g., blood, saliva, cell line, plasma, serum, tissue, formalin-fixed and/or paraffin-embedded samples, and whole genome-amplified DNA). In addition, the use of a universal biotinylated primer and multiplex analysis of up to three different amplicons can be performed, reducing genotyping cost and time of throughput. No other system provides this range, throughput, and cost advantage.

Within *Pyrosequencing® Protocols* the methods for utilizing Pyrosequencing technology are described in detail, including troubleshooting tips and background information. Chapter 1 provides an introduction to the fascinating origins of the methodology. Chapter 2 provides a brief overview of some of the applications Pyrosequencing® is used for. Chapters 3 and 4 describe primer selection and the basic technique. Chapters 5–7 provide methods for improving the throughput and decreasing the cost of Pyrosequencing®. Detailed applications for the technique can be found in Chapters 8–13, while the important aspect of data storage is discussed in Chapter 14.

The help and support of Rene Meyers and Robert England from Biotage AB has been invaluable, as has the assistance of Janine Wuebbles. In addition, I am indebted to all the authors for their hard work and outstanding contributions to *Pyrosequencing® Protocols*.

Sharon Marsh

Contents

Contributors

LEIF ANDERSSON • *Department of Medical Biochemistry and Microbiology, Uppsala University, Uppsala, Sweden*
PANAYIOTIS V. BENOS • *Department of Computational Biology, University of Pittsburgh School of Medicine, Pittsburgh, PA*
STEVEN C. ELBEIN • *Division of Endocrinology, Department of Internal Medicine, Central Arkansas Veterans Healthcare System and College of Medicine, University of Arkansas for Medical Sciences, Little Rock, AR*
MATTHEW J. ELLINGTON • *Antibiotic Resistance Monitoring and Reference Laboratory, Centre for Infections, Health Protection Agency, London, UK*
PAOLO FORTINA • *Department of Cancer Biology, Kimmel Cancer Center, Thomas Jefferson University, Philadelphia, PA and Dipartimento di Medicina Sperimentale e Patologia, Università degli Studi "La Sapienza," Rome, Italy*
DONG-CHUAN GUO • *Department of Internal Medicine, The University of Texas Medical School at Houston, Houston, TX*
IVO GLYNNE GUT • *Department of Technology Development, Centre National de Génotypage, Evry, France*
CRISTI R. KING • *Division of Oncology, Washington University School of Medicine, St. Louis, MO*
YING LU • *Division of Immunogenetics, Department of Pediatrics, Rangos Research Center, Children's Hospital of Pittsburgh, University of Pittsburgh School of Medicine, Pittsburgh, PA*
SHARON MARSH • *Division of Oncology, Washington University School of Medicine, St. Louis, MO*
DIANNA M. MILEWICZ • *Department of Internal Medicine, The University of Texas Medical School at Houston, Houston, TX*
SARAH E. NORTH • *Antibiotic Resistance Monitoring and Reference Laboratory, Centre for Infections, Health Protection Agency, London, UK*
PÅL NYRÉN • *Department of Biotechnology, AlbaNova University Center, Stockholm, Sweden*
PRITESH PATEL • *Department of Cancer Biology, Kimmel Cancer Center, Thomas Jefferson University, Philadelphia, PA*
CHRISTOPHER PECORARO • *Division of Immunogenetics, Department of Pediatrics, Rangos Research Center, Children's Hospital of Pittsburgh, University of Pittsburgh School of Medicine, Pittsburgh, PA*

Gerli Pielberg • *Department of Medical Biochemistry and Microbiology, Uppsala University, Uppsala, Sweden*
Ahad A. Rahim • *Gene Therapy Group, Chester Beatty Laboratories, Institute of Cancer Research, London, UK*
Steven Ringquist • *Division of Immunogenetics, Department of Pediatrics, Rangos Research Center, Children's Hospital of Pittsburgh, University of Pittsburgh School of Medicine, Pittsburgh, PA*
William A. Rudert • *Division of Immunogenetics, Department of Pediatrics, Rangos Research Center, Children's Hospital of Pittsburgh, University of Pittsburgh School of Medicine, Pittsburgh, PA*
Tiffany Scott-Horton • *Division of Oncology, Washington University School of Medicine, St. Louis, MO*
Yih-Horng Shiao • *Laboratory of Comparative Carcinogenesis, Center for Cancer Research, National Cancer Institute at Frederick, Frederick, MD*
Alexis Styche • *Division of Immunogenetics, Department of Pediatrics, Rangos Research Center, Children's Hospital of Pittsburgh, University of Pittsburgh School of Medicine, Pittsburgh, PA*
Jörg Tost • *Department of Technology Development, Centre National de Génotypage, Evry, France*
Massimo Trucco • *Division of Immunogenetics, Department of Pediatrics, Rangos Research Center, Children's Hospital of Pittsburgh, University of Pittsburgh School of Medicine, Pittsburgh, PA*
Derek J. Van Booven • *Division of Oncology, Washington University School of Medicine, St. Louis, MO*
Hua Wang • *Division of Endocrinology, Department of Internal Medicine, Central Arkansas Veterans Healthcare System and College of Medicine, University of Arkansas for Medical Sciences, Little Rock, AR*
Jonathon Wasson • *Department of Medicine, Washington University School of Medicine, St. Louis, MO*
Neil Woodford • *Antibiotic Resistance Monitoring and Reference Laboratory, Centre for Infections, Health Protection Agency, London, UK*

1

The History of Pyrosequencing®

Pål Nyrén

Summary

One late afternoon in the beginning of January 1986, bicycling from the lab over the hill to the small village of Fullbourn, the idea for an alternative DNA sequencing technique came to my mind. The basic concept was to follow the activity of DNA polymerase during nucleotide incorporation into a DNA strand by analyzing the pyrophosphate released during the process. Today, the technique is used in multidisciplinary fields in academic, clinical, and industrial settings all over the word. The technique can be used for both single-base sequencing and whole-genome sequencing, depending on the format used.

In this chapter I will give my personal account of the development of Pyrosequencing®—beginning on a winter day in 1986, when I first envisioned the method—until today, nearly 20 yr later. I will conclude with future prospects for the method.

Key Words: Pyrosequencing; sequencing; bioluminescence; pyrophosphate.

1. Introduction

Pyrosequencing® is a DNA sequencing technique that utilizes enzyme-coupled reactions and bioluminescence to monitor the pyrophosphate release accompanying nucleotide incorporation, in real-time. The Pyrosequencing method *(1,2)* is the first and only currently available commercial alternative to the well-known Sanger method for *de novo* DNA sequencing. Several hundred scientific papers comprise the literature describing the development and applications of Pyrosequencing (www.ncbi.nlm.nih.gov/gquery/gquery.fcgi; www.biotage.com). The method can be used for single-nucleotide polymorphism (SNP) analysis and tag sequencing (up to 100 bases), as well as for whole-genome sequencing *(3)*.

It was during my post-doctoral period in Cambridge when it occurred to me that detection of pyrophosphate could be used to sequence DNA. While working with Sir John Walker (later awarded the 1997 Nobel Prize for Chemistry)

From: *Methods in Molecular Biology, vol. 373: Pyrosequencing® Protocols*
Edited by: S. Marsh © Humana Press Inc., Totowa, NJ

at the Medical Research Council Laboratory of Molecular Biology (LMB), one of my projects was to isolate and sequence the gene for the bovine mitochondrial phosphate-carrier protein *(4)*. Sanger sequencing was performed manually, using radioactive-labeled nucleotides and handmade thin gels for detection and separation. Sequencing was an arduous business at the time, requiring weeks of practice to learn the procedure. Several steps were involved in the method, leaving time for the mind to wander in between. As a newcomer, I was not always successful with the handling of very thin acrylamide gels. I remember thinking that it would be great if the method could be simplified or if sequencing could be performed in some other way. During my time as a PhD student at Stockholm University, I worked extensively with modification and simplification of a variety of methods, as well as with the development of new procedures, so it was natural for me to look for ways to improve methods that I found cumbersome and labor intensive.

One late afternoon in the beginning of January 1986, bicycling from the lab over the hill to the small village of Fullbourn, the idea for an alternative DNA sequencing technique came to my mind. It was late, dark, and rainy as I hurried home to tell my wife Maija about the new idea. She later told me that when I explained the new idea to her, she thought that I looked like Gyro Gearloose's little helper—the bright-headed assistant with a light bulb as a head. I had difficulty sleeping that night and was eager to go home to Sweden to test my new idea. What I could not expect that day was that 10 yr would pass before the method was fully developed.

The basic concept was to follow the activity of DNA polymerase during nucleotide incorporation into a DNA strand by analyzing the pyrophosphate released during the process. Why pyrophosphate detection? As a PhD student I worked in the fields of bioenergetics and photosynthesis. My PhD project was to isolate and study an enzyme endogenous to a photosynthetic bacterium. The enzyme, proton-translocating inorganic pyrophosphatase, is involved in photophosphorylation and catalyzes light-driven pyrophosphate synthesis. In order to follow the light-driven process, I developed a very sensitive luminometric method *(5)* capable of following pyrophosphate synthesis in real-time, both after continuous illumination *(6)* and after short light flashes *(7)*. This was a break-through in bioenergetics research. The new idea was to follow the DNA polymerase activity using the pyrophosphate method, thereby detecting whether or not a base is incorporated during DNA synthesis. Because Watson–Crick base pairing rules that nucleotide G will always pair C and T with A, DNA polymerase activity will decipher the base composition of the template when known nucleotides are utilized.

As stated previously, it required nearly 10 yr to get the sequencing method working, involving much struggle to get financing and support for my idea, and to solve several problems that turned up as the project evolved. In the following sections, I will describe the most important parts of the developing process.

2. Envisioning Pyrosequencing

As I previously mentioned, I wanted to go back to Sweden as quickly as possible to test my new idea in the lab. However, we could not leave Cambridge before my post-doctoral period was completed. Of course I also wanted to accomplish something significant with my work at LMB before leaving. To expedite the process, I wrote a letter to my friend Åke Strid, asking him to advise a new student to purchase the necessary chemicals and set up a pilot study of the DNA polymerase assay. When I called after a month to check how things were going, he informed me that he could not get the suggested assay to work. I had to wait until the end of November 1986 before I could investigate what went wrong with my idea.

Back in Stockholm, I applied for research funding for two projects: my old bioenergetics project and the new sequencing concept. Unfortunately, I obtained financing only for the old project. The committee did not share my enthusiasm for the new idea, reasoning that ATP is a substrate for DNA synthesis and therefore would interfere with the luciferase assay. However, this point is wrong: ATP is not a substrate for DNA polymerase. The head of the department at that time had already advised me (without reading my project plans) not to apply for two separate projects because of the tough competition. Later, when I had already moved to the Royal Institute of Technology (KTH), and the project was shown to be successful, he alleged that it was a pity that I did not have the idea during my time at the Stockholm University.

Although I did not get research money for the DNA sequencing project, I decided to use a small portion of money from the other project to buy a few necessary chemicals and enzymes. I then spent evenings and weekends developing the DNA polymerase assay *(8)*. I eventually found that the method worked. I do not know why it had not worked for the student. I have learned that it is always better to do things yourself, if possible, rather then try to persuade another person, who might already be distracted by other projects. Even after the paper was published, I continued to apply for research funding without success. The project was shelved and I had to wait several years before I could continue to develop the method for use in DNA sequencing.

Much later, I learned that Bob Melamede *(9)*, whom I met in Stockholm in 1997, had described the general principles of DNA sequencing-by-synthesis in a previously obtained patent. The method was based on detection of the decrease in nucleotide absorbance upon nucleotide incorporation. The method's sensitivity was not sufficient for normal DNA concentrations. The pyrophosphate approach had an important advantage in terms of sensitivity; DNA quantities obtained by standard PCR procedures are adequate for analysis with this firefly-luciferase-based method. The sensitivity with absorbance–difference measurements is probably 100–1000 times lower. When I later met Bob, he was very happy to hear that his sequencing-by-synthesis concept worked and that I had circumvented the problem of DNA polymerase-activity monitoring.

3. Solid-Phase Pyrosequencing and the α-S-Nucleotide

In 1990, I moved to the KTH in Stockholm to join Professor Jan Rydstöm's group. Together with Åke Strid, studies of the bovine membrane-bound transhydrogenase were initiated, and we were also able to continue our old projects in Rydstöm's lab. As a newcomer at KTH, I spent some time reading old theses and articles published by the Biochemistry department. One especially attracted my interest: it described a method for solid-phase DNA sequencing ***(10)***. Paramagnetic beads were utilized for DNA template preparation before sequencing. I theorized that if I could combine the magnetic bead technique with my DNA analysis system, a DNA sequencing procedure would be possible. I consulted Professor Mathias Uhlén, who was the principle investigator for the magnetic bead project, and shortly thereafter I collaborated with Bertil Pettersson, a student from Mathias Uhlén's group. Bertil taught me everything that I needed to know about working with magnetic beads. He synthesized oligonucleotides and prepared DNA by PCR. I worked with the magnetic beads and the pyrophosphate detection system. During the first 4 yr I worked 1–2 d a week with the new project.

A problem I faced during the initial stages of the project was high background during the luminescence measurement. The enzymes and the nucleotides were the source of most of the background signals. One approach I used to lower the background was preincubation with apyrase immobilized on magnetic beads; the method was published several years later ***(11)***. I also used to pretreat the nucleotides with pyrophosphatase to decrease the pyrophosphate content. Today, this is a standard procedure. The most severe problem I observed was that dATP functioned as a weak substrate for luciferase giving light corresponding to 2–3% of an equal amount of ATP. I mediated this difficulty by starting all reactions by adding DNA polymerase and not by adding nucleotides. This procedure made it unnecessary to subtract a high background from a low signal. Because most polymerases were delivered in phosphate buffers, these enzymes also had to be pretreated with pyrophosphatase.

Although solutions to several of the mentioned problems had now been found, I still had difficulty getting strong, clear signals. Bertil then came up with the brilliant proposal of setting up the analysis system quite differently.

We used run-off signals to increase signal intensity, and dideoxynucleotides were used to facilitate reading the first base position. This new method was the first proof-of-principle for the DNA sequencing concept, and the success was encouraging for future developmental work ***(12)***. Later, we published an alternative method for single-base change detection ***(13)***. The concept relies on measuring the differences in primer extension efficiency by a DNA polymerase of a matched over a mismatched 3'-terminal utilizing α-thiotriphosphate analogs. One important feature of the DNA sequencing concept is that DNA polymerases lacking 3'–5' activity are required, which limited us to the use of Sequenase and a modified Klenow enzyme. When we used dideoxynucleotides, we could only use Sequenase because of their hindrance to Klenow's activity.

To be able to sequence multiple bases, we had to work on decreasing the dATP background. I made a literature search for alternative nucleotides and decided to test an α-*S*-modified nucleotide. Nucleoside thiophosphates comprised a new class of modified nucleotides in which one nonbridging oxygen atom in the α-phosphate of the nucleoside 5'-triphosphate is replaced by a sulfur group. I found that the modified nucleotide was a good substrate for the polymerase and in addition, a poor substrate for luciferase. This finding was made during 1995 and was published 1996 ***(14)***. We were able to sequence 15 bases with our new approach—a new world record for the sequencing-by-synthesis principle.

From 1991 to 1994 I was engaged in several other projects ***(15–17)***, and the development speed for the DNA sequencing method was low. It was essentially a one-man project with little or no funding, and I spent about 25% of my time with research and the rest with teaching obligations. From 1994 to 1998 my one-man group increased to several people, at most eight, working with different aspects of the DNA method. No one was particularly enthusiastic about the new idea during the earlier period—when I presented my idea at a conference in Stockholm very few people showed interest in my poster. However, one well-known scientist (an adviser for the government) approached my poster, looked it over, and then turned to me, saying, "I don't think this will ever work." That comment made me a bit anxious, as she was an expert in nucleotide metabolism. Regardless, I continued to believe in my idea.

Up to this point, we had used magnetic beads and manual sequencing, but our aim was to automate the procedure. Together with Professor Johan Roeraade, PhD, Sean Waters and my students Mostafa Ronaghi, Tommy Nordström, and Atefeh Shakri, I started to explore the possibility of immobilizing enzymes and DNA on silica with the goal of constructing an automated sequencing procedure utilizing a capillary flow system. At that time, we also started a collaboration with Björn Ekström at Pharmacia Biotech. We were able to immobilize both luciferase and ATP sulfurylase and successfully analyze ATP and pyrophosphate continuously in the capillary system (unpublished results). We also made preliminary studies of nucleotide incorporations on immobilized DNA.

4. Apyrase

While attempting to solve problems such as low-signal intensity and nucleotide contamination associated with the capillary flow system, another idea occurred to me. Instead of including a washing step between the nucleotide additions, it might be possible to utilize a nucleotide-degrading enzyme. My first thought was to use my earlier published ***(11)*** concept of apyrase immobilized on magnetic beads. The immobilized apyrase could be separated from the assay after each degradation step by simply using a magnet. Alternatively, if the kinetics were properly adjusted, it might be possible to omit the separation step altogether. However, it remained to be seen whether apyrase could degrade nucleotides other than ATP and if all four deoxynucleotides were equally well degraded. At that time, no one was especially impressed or excited with this proposal. I asked one of my students to test the concept, but he later told me that it did not work. I could not let the idea go, so during the summer 1996 vacancies I went back to the lab bench and started to set up some really exciting experiments.

After about 6 wk in the lab I had preliminary data for a new DNA sequencing procedure ready. The main obstacle that I had to handle was false signals, which I theorized were a result of nucleoside diphosphate kinase activity. I identified the ATP sulfurylase preparation as the main contamination source for this activity. As predicted, the false signals decreased with lower concentrations of ATP sulfurylase. Two months later I had evidence supporting my theory ***(18)***. I started using a new source of ATP sulfurylase purified from dry yeast (bought from a nearby food store) by my student Nader Nourizad. Another student, Samer Karamohamed, was later able to produce a recombinant form of the enzyme in *Escherichia coli* ***(19)***. Both the purified enzyme and the recombinant ATP sulfurylase improved the sequencing results dramatically. Together with Ronaghi, I started to optimize all parameters of the method, and in early 1998 the protocol was ready ***(1,2)***. Only minor changes have since been made to the protocol.

One problem we encountered during the optimization process was that some templates produced significantly better results than others. We hypothesized that some DNA templates formed secondary structures because of the relatively low temperature (22°C) that was used. We tried to solve this issue by adding different substances such as glycerol, proline, and DMSO, but none of these cheaper substances helped with this problem. Ronaghi then suggested that we should test the effects of adding a single-stranded DNA-binding protein (SSB). Most organisms utilize this protein to decrease secondary-structure formation, thereby improving DNA synthesis. I had earlier found that SSB could be used to improve DNA priming with two or three unligated hexamers, so the substance was still available in our lab. SSB was a hit; it dramatically improved the sequence quality for difficult templates and for less difficult templates to some degree as well ***(20,21)***.

5. Automation

With the aim of commercializing the new DNA sequencing technique, the company Pyrosequencing AB (today Biotage AB) was founded in 1997 by Pål Nyrén, Mathias Uhlén, Mostafa Ronaghi, Bertil Pettersson, and Björn Ekström. The first commercially available automated system was sold 1999. The first noncommercially available automated system was developed by myself, Tommy Nordström, and Mostafa Ronaghi ***(22)***. It included an LKB-1251 luminometer, two dispenser controllers, a power unit, a computer, a recorder, and four separate dispensers. At 1–5 μL, the dispensed volume was rather large, as was the sample volume at 0.4 mL. The system was also restricted to sequencing one sample at a time. In contrast, the commercial systems developed later sequence 96 samples in parallel, and both dispensation volumes (0.05–0.2 μL) and sample volumes (10–50 μL) are much smaller, substantially decreasing the cost and time for a DNA sequencing project.

6. Improvements

In addition to the development process, several aspects of the Pyrosequencing method have been improved during the last few years. I will briefly mention a few of the most important alterations. We had observed major decreases in apyrase activity following the substitution of an S-modified variant for the standard dATP during longer sequencing projects. The new nucleotide used was a mixture of two isomers—one of which, the R-isomer, is not a substrate for the DNA polymerase. We therefore decided to test a nucleotide solution consisting of only the active S-isomer. By utilizing a pure isomer, we could decrease the nucleotide concentration by one-half and thereby dramatically decrease the apyrase inhibition. We obtained much longer reads using the new approach ***(23)***; up to 153 bases of sequence information could be analyzed on one of the studied templates.

It was obvious from the previously mentioned experiment that the modified A nucleotide had a negative effect on apyrase activity. Although the problem was decreased by the use of the pure isomer we could still observe apyrase inhibition, especially during long sequencing projects. After a deep literature search, I came to the conclusion that only a few nucleotide analogs were potentially appropriate for the Pyrosequencing method. We tested 7-deaza-2'-

deoxyadenosine-5'-triphosphate and observed positive effects on the sequencing data. The nucleotide was incorporated by the polymerase with good efficiency, and in addition it had no negative effect on apyrase activity even after many nucleotide additions *(24)*. We surmise that this nucleotide might be a good choice for use in future Pyrosequencing protocols.

Another improvement of the sequencing method, especially for templates with homopolymeric T-regions, has been the introduction of Sequenase. For example, a template containing an 8-mer poly(T) track could be easily read when Sequenase was used, but not when Klenow polymerase was used *(25)*. Sequenase could also be used in combination with 7-deaza-2'-deoxyadenosine-5'-triphosphate *(24)*.

We had showed earlier that SSB improved the sequence quality for some templates but the effect of temperature had not been investigated because of the thermo-instability of firefly luciferase. By introducing glycine betaine we were able to increase the temperature for the sequencing procedure by approx 10°C, up to 37°C *(26,27)*. At this higher temperature, the activity of the different enzymes was doubled and the sequence quality for some templates improved.

Template preparation for Pyrosequencing has been substantially improved by the introduction of a Sepharose bead/vacuum system that allows preparation of 96 samples within a few minutes. Another strategy, introduced by Tommy Nordström, allows double-stranded DNA to be used as a template for DNA sequencing after a simple preparation procedure *(28–30)*. Using a combination of apyrase, pyrophosphatase, and blocking oligonucleotides, the template can be prepared by a one-step procedure within a few minutes *(30)*.

Preprogrammed dispensation order was introduced to improve read length *(31)*. This strategy has applications in resequencing projects, such as mutation detection in cancer research and clone checking *(31,32)*.

A major improvement of the Pyrosequencing method was the introduction of the multiprimer DNA sequencing concept by my student Baback Gharizadeh *(33–35)*. In this method, two or more sequencing primers, combined in a pool, are added to a DNA sample of interest. The oligonucleotides that hybridize to the DNA sample will function as primers during the subsequent DNA sequencing procedure. This new strategy is suitable for sequencing and typing of samples harboring different genotypes (coinfections with multiple genotypes) and samples yielding nonspecific amplifications. This therefore eliminates the need for nested PCR, stringent PCR conditions, and cloning. The new approach has also proved to be useful for amplicons containing low yields or subdominant types.

7. Applications

DNA sequencing has become an invaluable tool in such disparate fields as medicine, agriculture, and forensic studies. Pyrosequencing can be used for both single-base sequencing and whole-genome sequencing, depending on the format used. The recently developed picotiterplate format allows whole-genome sequencing with very high throughput: a 100-fold increase over current Sanger sequencing technology ***(3,36)***. The conventionally used microtiterplate format, developed by Biotage AB (www.biotage.com), is used to sequence 96 samples in parallel for up to or more than 100 bases ***(23,34)***. Earlier used primarily for SNP analyses ***(28,37)***, the method now has numerous different applications: analysis of allele frequency in pooled samples ***(38)***, methylation analyses ***(39)***, molecular haplotyping ***(40)***, sequencing of heteroplasmic DNA ***(41)***, forensic analyses ***(42–44)***, bacterial typing ***(34)***, fungal typing ***(45)***, and viral typing ***(46)***. Furthermore, Pyrosequencing facilitates clinical research in areas such as Alzheimer's disease, autoimmune disorders, bowel disease, cardiovascular disease, coronary heart disease, dermatology, diabetes, gynecology, hematology, hearing loss, hematopoietic chimerism, immunology, mitochondrial disorders, nephrology, neurology, obesity, oncology, orthopedics, psychiatric genetics, and trauma (for references *see* www.ncbi.nlm.nih.gov/gquery/gquery.fcgi; www.biotage.com). The method is also used for animal ***(47)*** and plant studies ***(48)***.

8. Future Prospects

With the completion of the human genome-sequencing project, there is now a focus on developing rapid new methodologies that will enable routine genomics studies and molecular testing in clinical settings. The future prospects for Pyrosequencing are encouraging, but present numerous challenging possibilities. The method can be used in many different formats, from single-sample analysis to analysis of millions of samples in parallel, utilizing volumes ranging from the milliliter-to-picoliter scale. Pyrosequencing is now established as a very reliable method for diagnostic sequencing of short DNA stretches. The information it provides about sequence context for SNP analysis makes it the most robust method available at this time. The procedure has also been shown to be appropriate for whole-genome sequencing with very high throughput. Increasing read length is among the greatest challenges related to Pyrosquencing. In forensics, being able to analyze short tandem repeat markers by sequencing might help to overcome problems encountered with traditional size determination methods.

The Pyrosequencing method relies on the cooperation of four different enzymatic reactions. One way to aid understanding of this complicated process is to use mathematical models. Once a reliable model has been developed, it can be used to find ways of improving the Pyrosequencing process itself. This could, for instance, involve modifications of the substrate concentrations, or use of other enzymes with different kinetic properties. Performing full tests like these in the laboratory is often both expensive and time-consuming, which is why an *in silico* model may help to identify especially promising configurations. The model would be designed to find the parameters producing an optimal sequencing result given certain input criteria.

We have recently constructed a mathematical model of the Pyrosequencing reaction system utilizing the assumptions of irreversible Michaelis–Menten rate equations ***(49)***. In this first published model the dynamic of a single light pulse was captured with great accuracy, as well as the overall characteristics of a whole Pyrogram®. The simulation results of the mathematical model show significant potential, motivating further development of the model for detail and accuracy. The approaching task is to find a way to replace the artificial factors of incorporation efficiency and plus-shifts with a nonfixed variable. Incorporation efficiency should, for instance, be dependent on polymerase velocity, nucleotide concentration, time, and quantity of complementary DNA. In order to include these parameters in the first model, a more sophisticated description of the polymerase's kinetics is likely needed. Another factor is a polymerase's processivity—the number of nucleotides the enzyme can incorporate before it releases them from the DNA strand. For the Klenow fragment, the processivity is approx 20 when the enzyme is saturated by nucleotides. The first model utilizes a processivity of one. With these features incorporated into the model, information about the benefits and drawbacks of using a different polymerase, with other kinetic properties, can be investigated. We believe that there is great potential for increasing Pyrosequencing's read length, thereby extending the method's applicability in new and different fields.

References

1. Nyrén, P. (2001) Method for sequencing DNA based on the detection of the relese of pyrophosphate and enzymatic nucleotide degradation. Patents: US 6 258 568BI and WO98/28440.
2. Ronaghi, M., Uhlén, M., and Nyrén, P. (1998) A sequencing method based on real-time pyrophosphate detection. *Science* **281,** 363–365.
3. Margulies, M., Egholm, M., Altman, W. E., et al. (2005) Genome sequencing in microfabricated high-density picolitre reactors. *Nature* **437,** 376–380.
4. Runswick, M. J., Powell, S. J., Nyrén, P., and Walker, J. W. (1987) Sequence of the bovine mitochondrial phosphate carrier protein: structural relationship to ADP/

ATP translocase and the brown fat mitochondria uncoupling protein. *EMBO J.* **6,** 1367–1373.

5. Nyrén, P. and Lundin, A. (1985) Enzymatic method for continuous monitoring of inorganic pyrophosphate synthesis. *Anal. Biochem.* **151,** 504–509.
6. Nyrén, P., Nore, B. F., and Baltscheffsky, M. (1986) Studies on photosynthetic inorganic pyrophosphate formation in *Rhodospirillum rubrum* chromatophores. *Biochim. Biophys. Acta* **851,** 276–282.
7. Nyrén, P., Nore, B. F., and Baltscheffsky, M. (1986) Inorganic pyrophosphate synthesis after a short light flash in chromatophores from *Rhodospirillum rubrum. Photobiochem. Photobiophys.* **11,** 189–196.
8. Nyrén, P. (1987) Enzymatic method for continuous monitoring of DNA-polymerase activity. *Anal. Biochem.* **167,** 235–238.
9. Melamede, R. J. (1985) Automatable process for sequencing nucleotide. US Patent 4863849.
10. Ståhl, S., Hultman, T., Moks, T., and Uhlén, M. (1988) Solid phase DNA sequencing using the biotin-avidin system. *Nucleic Acids Res.* **16,** 3025–3038.
11. Nyrén, P. (1994) Apyrase immobilized on paramagnetic beads used to improve detection limits in bioluminometric ATP monitoring. *J. Biolumin. Chemilumin.* **9,** 29–34.
12. Nyrén, P., Pettersson, B., and Uhlén, M. (1993) Solid phase DNA minisequencing by an enzymatic luminometric inorganic pyrophosphate detection assay. *Anal. Biochem.* **208,** 171–175.
13. Nyrén, P., Karamouhamed, S., and Ronaghi, M. (1997) Detection of single-base changes using a bioluminometric primer extension assay. *Anal. Biochem.* **244,** 367–373.
14. Ronaghi, M., Karamohamed, S., Pettersson, B., Uhlén, M., and Nyrén, P. (1996) Real-time DNA sequencing using detection of pyrophosphate release. *Anal. Biochem.* **242,** 84–89.
15. Nyrén, P. (1994) A method for the detection of cells, cell lysis or cell-lysing activity, Swedish patent application.
16. Nyrén, P. and Edwin, V. (1994) Inorganic pyrophosphate-based detections: Detection and enumeration of cells. *Anal. Biochem.* **220,** 39–45.
17. Nyrén, P. and Edwin, V. (1994) Inorganic pyrophosphate-based detections: Detection and quantification of cell lysis and cell-lysing activity. *Anal. Biochem.* **220,** 46–52.
18. Karamohamed, S., Nordström, T., and Nyrén, P. (1999) A real-time bioluminometric method for detection of nucleoside diphosphate kinase activity. *Biotechniques* **26,** 728–734.
19. Karamouhamed, S., Nilsson, J., Nourizad, K., Ronaghi, M., Pettersson, B., and Nyrén, P. (1999) Production, purification, and real-time functional analysis of recombinant *Saccharomyces cervisiae* MET3 adenosine triphosphate sulfurylase expressed in *Escherichia coli. Protein Expr. Purif.* **15,** 381–388.
20. Ronaghi, M. (2000) Improved performance of Pyrosequencing using single-stranded DNA-binding protein. *Anal. Biochem.* **286,** 282–288.

21. Nordström, T., Gharizadeh, B., Pourmand, N., Nyrén, P., and Ronaghi. M. (2001) Method enabling fast partial sequencing of cDNA clones. *Anal. Biochem.* **292,** 266–271.
22. Nordström, T., Ronaghi, M., and Nyrén, P. (1999) Automation of a novel DNA sequencing method. In: *Bioluminescence and Chemiluminescence: Perspective for the 21st Century,* (Roda, A., Pazzagli, M., Kricka, L. J., and Stanley, P. E., eds.), John Wiley and Sons, Hoboken, NJ, pp. 528–531.
23. Gharizadeh, B., Nordström, T., Ahmadian, A., Ronaghi, M., and Nyrén, P. (2002) Long read Pyrosequencing using pure 2'-deoxyadenosine-5'-O'-(1-thiotriphosphate) Sp-isomer. *Anal. Biochem.* **301,** 82–90.
24. Eriksson, J., Gharizadeh, B., Nourizad, N., and Nyrén, P. (2004) 7-deaza-2'-deoxyadenosine-5'-triphosphate as an alternative nucleotide for the Pyrosequencing technology. *Nucleosides Nucleotides Nucleic Acids* **23,** 1583–1594.
25. Gharizadeh, B., Eriksson, J., Nourizad, N., Nordström, T., and Nyrén, P. (2004) Improvements in Pyrosequencing technology by employing Sequenase polymerase. *Anal. Biochem.* **330,** 272–280.
26. Eriksson, J., Nordström, T., and Nyrén, P. (2003) Method enabling firefly luciferase based bioluminometric assays at elevated temperature. *Anal. Biochem.* **314,** 158–161.
27. Eriksson, J., Gharizadeh, B., Nordström, T., and Nyrén, P. (2004) Pyrosequencing technology at elevated temperature. *Electrophoresis* **25,** 20–27.
28. Nordström, T., Ronaghi, M., Morgenstern, R., Ekström, L., de Faire, U., and Nyrén, P. (2000) Direct analysis of single nucleotide polymorphism on double-stranded DNA. *Biotechnol. Appl. Biochem.* **31,** 107–112.
29. Nordström, T., Nourizad, K., Ronaghi, M., and Nyrén, P. (2000) Method enabling Pyrosequencing on double-stranded DNA. *Anal. Biochem.* **282,** 186–193.
30. Nordström, T., Alderborn, A., and Nyrén, P. (2002) Method for one-step preparation of double-stranded DNA template applicable for use with Pyrosequencing technology. *J. Biochem. Biophys. Methods* **52,** 71–82.
31. Garcia, C. A., Ahmadian, A., Garizadeh, B., Lundeberg, J., Ronaghi, M., and Nyrén, P. (2000) Mutation detection by Pyrosequencing: sequencing of exons 5 to 8 of the p53 tumor suppressor gene. *Gene* **253,** 249–257.
32. Nourizad, N., Gharizadeh, B., and Nyrén, P. (2003) Method for clone checking. *Electrophoresis* **24,** 1712–1715.
33. Gharizadeh, B., Ghaderi, M., Donnelly, D., Wallin, K-L., and Nyrén, P. (2003) Multiple-primer DNA sequencing method. *Electrophoresis* **24,** 1145–1151.
34. Gharizadeh, B., Ohlin, A., Mölling, P., et al. (2003) Multiple group-specific sequencing primers for reliable and rapid DNA sequencing. *Mol. Cell. Probe* **17,** 203–210.
35. Gharizadeh, B., Oggionni, M., Zheng, B., et al. (2005) Type-specific multiple sequencing primers: a novel strategy for reliable and rapid genotyping of human papillomaviruses by Pyrosequencing technology. *J. Mol. Diagn.* **7,** 198–205.
36. Leamon, J. H., Lee, W. L., Tartaro, K. R., et al. (2003) A massively parallel PicoTiterPlate based platform for discrete picoliter-scale polymerase chain reaction. *Electrophoresis* **24,** 3769–3777.

37. Ahmadian, A., Gharizadeh, B., Gustafsson, A., et al. (2000) Single nucleotide polymorphism analysis by Pyrosequencing. *Anal. Biochem.* **280,** 103–110.
38. Gruber, J. D., Colligan, P. B., and Wolford, J. K. (2002) Estimation of single nucleotide polymorphism allele frequency in DNA pools by using Pyrosequencing. *Hum. Genet.* **110** , 395–401.
39. Uhlmann, K., Brinckmann, A., Toliat, M. R., Ritter, H., and Nurnberg, P. (2002) Evaluation of a potential epigenetic biomarker by quantitative methyl-single nucleotide polymorphism analysis. *Electrophoresis* **23,** 4072–4079.
40. Ahmadian, A., Lundeberg J., Nyrén, P., Uhlén, M., and Ronaghi, M. (2000) Analysis of the p53 tumor supressor gene by Pyrosequencing. *Biotechniques* **28,** 140-–147.
41. Goriely, A, McVean, G. A., Rojmyr, M., Ingemarsson, B., and Wilkie, A.O. (2003) Evidence for selective advantage of pathogenic FGFR2 mutations in the male germ line. *Science* **301,** 643–646.
42. Andreasson, H., Asp, A., Alderborn, A., Gyllensten, U., and Allen, M. (2002) Mitochondrial sequence analysis for forensic identification using Pyrosequencing technology. *Biotechniques* **32,** 124–133.
43. Allen, M. and Andreasson, H. (2005) Mitochondrial D-loop and coding sequence analysis using Pyrosequencing. *Methods Mol. Biol.* **297,** 179–196.
44. Cebula, T. A., Brown, E. W., Jackson, S. A., Mammel, M. K., Mukherjee, A., and LeClerc, J. E. (2005) Molecular applications for identifying microbial pathogens in the post-9/11 era. *Expert Rev. Mol. Diagn.* **5,** 431–445.
45. Gharizadeh, B., Norberg, E., Löffler, J., et al. (2004) Identification of medically important fungi by Pyrosequencing technology. *Mycoses* **47,** 29–33.
46. Gharizadeh, B., Kalantari, M., Garcia, C., Johansson, B., and Nyrén, P. (2001) Typing of human papillomavirus (HPV) by Pyrosequencing. *Laboratory Investigation* **81,** 673–679.
47. Milan, D., Jeon, J. T., Looft, C., et al. (2000) A mutation in PRKAG3 associated with excess glycogen content in pig skeletal muscle. *Science* **19,** 1248–1251.
48. Mochida, K., Yamazaki, Y., and Ogihara, Y. (2003) Discrimination of homoeologous gene expression in hexaploid wheat by SNP analysis of contigs grouped from a large number of expressed sequence tags. *Mol. Genet. Genomics* **270** , 371–377.
49. Svantesson, S., Westermark, P. O., Hellgren Kotaleski, J., Gharizadeh, B., Lansner, A., and Nyrén, P. (2004) A mathematical model of the Pyrosequencing reaction system. *Biophys. Chemist.* **110,** 129–145.

2

Pyrosequencing® Applications

Sharon Marsh

Summary

Genetics research has benefited tremendously from the release of the human genome sequence. Subsequent technology has been developed and adapted to accommodate the need for faster, easier throughput of genetic assays. Pyrosequencing® is a unique system that allows the analysis of genetic variations including single-nucleotide polymorphisms, indels and short repeats, as well as assessing RNA allelic imbalance, DNA methylation status, and gene copy number. Advances in methodology, including multiplex and universal primer applications, have reduced assay cost and improved throughput. This chapter briefly reviews some of the many applications for Pyrosequencing technology.

Key Words: Pyrosequencing; DNA; RNA; genetics.

1. Introduction

Since the release of a working draft of the human genome there has been a drive for technological advancements to improve the throughput and analysis of genetics research.

Low-throughput procedures such as restriction fragment length polymorphism and allele-specific PCR *(1)* are time consuming and laborious when multiple samples and/or multiple assays are to be used. Conversely, high-throughput procedures utilizing chips or bead arrays *(2)* are cheap per variant but are limited by the large amounts of sample needed and are consequently limited to studies such as whole-genome scanning where data on up to 500,000 polymorphisms are required. A range of medium-throughput technologies have emerged to fill the gap, each with their advantages and disadvantages *(3)*. Pyrosequencing is one such technology that allows rapid and reliable genotyping to be performed in a 96-well plate format.

From: *Methods in Molecular Biology, vol. 373: Pyrosequencing® Protocols*
Edited by: S. Marsh © Humana Press Inc., Totowa, NJ

2. Principles of Pyrosequencing

The origins of Pyrosequencing have been described in Chapter 1. This innovative technique is based on sequencing by synthesis. The assay takes advantage of the natural release of pyrophosphate whenever a nucleotide is incorporated onto an open 3' DNA strand. The released pyrophosphate is used in a sulfurylase reaction releasing ATP. The released ATP can be used by luciferase in the conversion of luciferin to oxyluciferin. The reaction results in the emission of light, which is collected by a CCD camera and recorded in the form of peaks, known as a Pyrogram® *(4–6)* (**Fig. 1**). When a nucleotide is not incorporated into the reaction, no pyrophosphate is released and the unused nucleotide is removed from the system by degradation through apyrase. This four-enzyme process is performed in a closed system in a single well. The processing time is simple and relatively fast (~20 min/96-well plate) and the cost of the reaction is comparable to other medium-throughput technologies.

3. Applications

The advantages with the Pyosequencing system lie with the range of applications the technology can be applied to (**Fig. 2**). Single-nucleotide polymorphisms (SNP), insertion/deletions (indels), short tandem repeats, pooled allele frequencies, human leukocyte antigen (HLA) typing, gene copy number, allelic imbalance in RNA, methylation status, and short sequencing stretches are among the numerous applications, many of which are described in this book. As short stretches of sequence are synthesized during the assay, novel polymorphisms close to the polymorphism in question have also been identified using this technique, where they may be missed or cause inaccurate genotype calls using other methods *(7)*. The assay is applicable to almost any source of DNA or RNA (e.g., blood, saliva, cell line, plasma, serum, tissue, formalin-fixed and/or paraffin-embedded samples, and whole genome-amplified DNA). In addition, the use of a universal biotinylated primer and multiplex analysis of up to three different amplicons can be performed *(8–10)*, reducing genotyping cost and time of throughput. No other system provides this range, throughput, and cost advantage. An up-to-date listing of publications involving Pyrosequencing technology can be found at http://www.biotagebio.com/DynPage.aspx?id=8890&search=publications.

3.1. Human Genetics

Historically, genetics research works back from a phenotype using family inheritance patterns and linkage analysis to narrow down regions associated with genetic disorders. However, once the genome regions have been assigned they often still span huge stretches of DNA and narrowing down further to identify the causative genes/haplotypes/variants can be an arduous process.

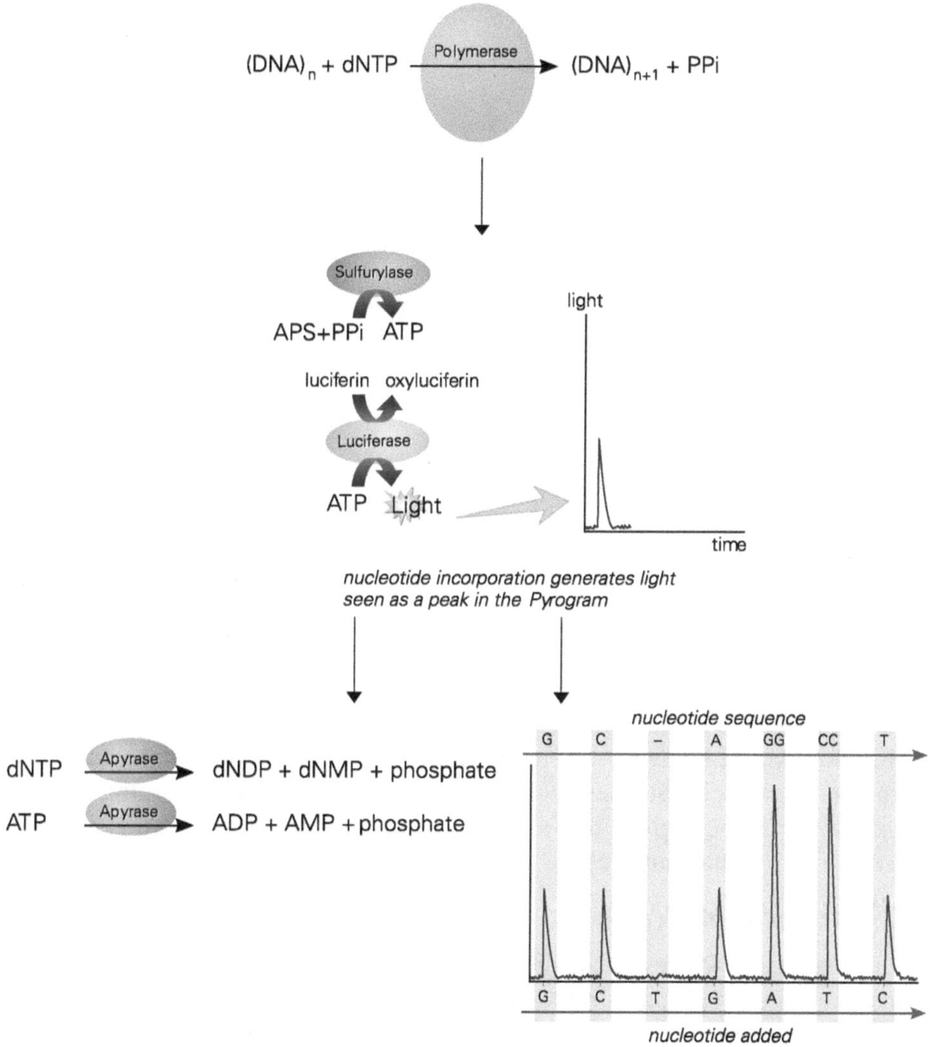

Fig. 1. The principles of Pyrosequencing®. (Reproduced with permission from **ref. *46*.**)

Utilizing the Pyrosequencing allele quantification software, pooling individual samples allows a rapid and cost-effective process for screening allele frequencies in affected and control populations ***(11)***. For example, Permutt et al. used a case–control design to screen 91 SNPs from a 7.3-Mb region in pooled samples of 150 individuals with diabetes and 150 controls from the same population (Ashkenazi) ***(12)***. This allowed a direct comparison of allele frequencies between cases and controls and was performed in 182 assays rather than the

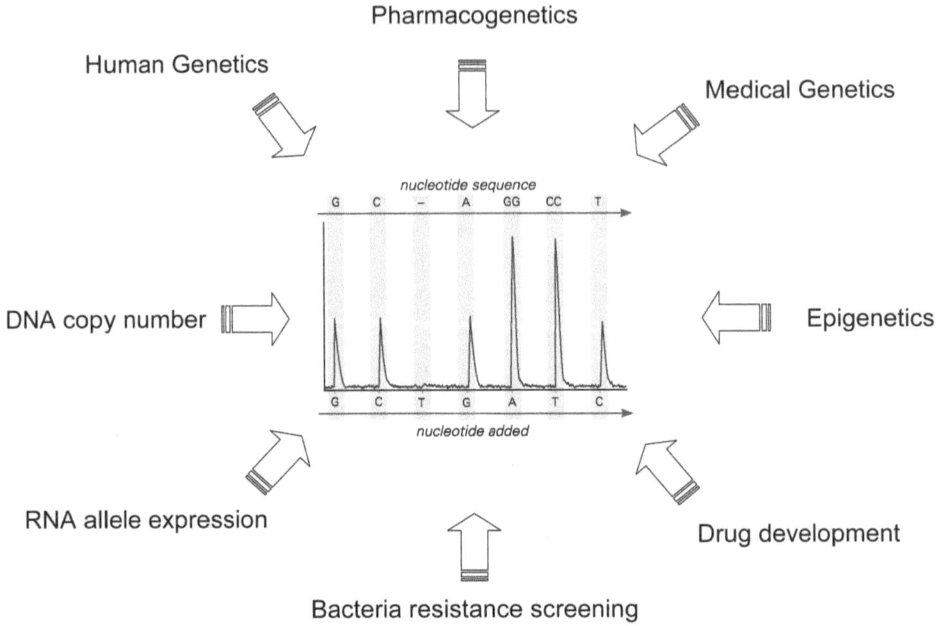

Fig. 2. Applications of Pyrosequencing®. Pyrogram® adapted from **Fig. 1**.

27,300 that would be necessary to screen all of the individuals for each polymorphism separately. Studies such as this can allow a rapid screening process to narrow down the regions that should be subsequently genotyped at a higher density in individual samples.

3.2. Pharmacogenetics

Pharmacogenetics research, the search for inherited variability in drug response, relies on rapid, accurate genotyping technologies. There are many examples of the use of Pyrosequencing for pharmacogenetics research in the recent literature *(13–27)*. Although the majority of DNA variation is in the form of individual DNA nucleotide differences (SNPs), other variants including tri-allelic polymorphisms, tandem repeats, and indels have been shown to be clinically relevant. For example, a functional dinucleotide (TA) repeat in the *UGT1A1* gene has been associated with severe toxicity in cancer patients treated with the chemotherapy drug, irinotecan *(28)*. The majority of people have six copies of the TA repeat. The *UGT1A1*28* polymorphism corresponds to seven copies of the repeat, and this is the allele associated with toxicity. This polymorphism can be identified by direct sequencing, but the analysis is time consuming and the assay can be expensive. An accurate Pyrosequencing assay has been developed that screens this polymorphism in a 96-well plate format and significantly reduces analysis time ***(24)***.

The ability to multiplex reactions is also a boon to pharmacogenetics research. The multidrug transporter gene, *ABCB1* (MDR1) has three commonly studied polymorphisms. As these are all usually screened in patient samples, assessing all three in one multiplexed assay ***(29)*** can reduce time, cost, and importantly, reduces the amount of patient DNA required.

Copy-number polymorphisms can also have pharmacogenetic relevance. CYP2D6 is another metabolizer of commonly used drugs. Gene duplication events lead to polymorphic *CYP2D6* gene copy numbers between individuals. Amplification of the *CYP2D6* gene is associated with increased enzyme levels and consequently increased CYP2D6-mediated metabolism. Time-consuming and labor-intensive methods such as long-range PCR can be used to quantify the number of *CYP2D6* genes in a patient. However, the development of a Pyrosequencing assay (Chapter 12) to quantify gene copy numbers ***(30)*** allows rapid and reliable data to be generated. For *CYP2D6* the assay takes advantage of a known pseudogene (*CYP2D8P*) and the allele quantification software to compare the ratio of peak heights and determine the number of *CYP2D6* genes present ***(31)***.

Along with identifying clinically relevant polymorphisms, an added complication can be found where an imbalance in allele expression is seen in the RNA expression of a gene. For example, in patients heterozygous for a functional polymorphism, if the deleterious allele is over-represented in the RNA compared with the wild-type allele, this could significantly affect patient outcome. Over-representation of specific alleles has been demonstrated in *CYP3A5*, a gene responsible for the metabolism of many commonly prescribed drugs. A polymorphism in the 3'-untranslated region of the *CYP3A5* gene is over-represented in cDNA from individuals whose DNA is heterozygous for the polymorphism ***(32)***. This is ultimately owing to tight linkage with a SNP in the promoter region of the gene causing increased expression ***(32)***. Consequently, the need to look to RNA for allele expression may provide more information than simply assessing the DNA for the presence of the functional allele. Chapter 13 details a method for assessing allelic imbalance, which could be readily applied to studies such as these. Consequently, assessment of RNA allele expression and DNA polymorphism in the status of the same individual can be readily performed on the same technology platform.

3.3. Epigenetics

Hypermethylation of CpG islands located within or close to the 5' region of genes is associated with inhibition of gene expression. This can be particularly important in tumors where hypermethylation can "switch off" tumor-suppressor genes and/or DNA damage repair genes, which can lead to sensitivity or resistance to chemotherapy drugs ***(33)***. Assessment of methylation status in CpG islands can be easily performed using bisulfate modification and comparing the conversion of cytidine to thymidine with the allele quantification Pyrosequencing software (Chapter 8). The procedure for detecting methylation using Pyrosequencing ***(34)*** was recently applied to determine the methylation status in oral squamous cell carcinoma ***(35)***, leading to the identification of *cytoglobin* as a novel putative tumor suppressor gene ***(35)***.

3.4. Medical Genetics

The use of Pyrosequencing to rapidly and accurately type HLA loci (Chapter 10) is of valuable clinical importance ***(36–41)***. The matching of patient and donor for HLA significantly reduces the chance of rejection following organ transplantation.

Other medical applications for Pyrosequencing include the typing of mutations (Chapter 9) conferring antibiotic resistance to bacteria or to differentiate between multiple bacterial or viral strains ***(42–44)*** and for screening for target phage ligands for novel drug development (Chapter 11) ***(45)***. These protocols demonstrate the utility of Pyrosequencing for clinical assessment and drug development applications.

The technique also has far-reaching applications to fields outside of human/medical genetics. Chapter 12 describes the use of gene copy number detection in animal studies where Pyrosequencing technology can be utilized to determine X and Y copy numbers in cattle sperm sorting, or coat color variation gene copies in pigs, demonstrating that the same technique and protocols can be applied to a wide variety of disciplines.

3.5. Conclusion

The availability of a multipurpose instrument for genetic analysis is a major boon to researchers. Pyrosequencing is an established, reliable technique that can be applied to all areas of genetics. Applications of Pyrosequencing technology can be found in widely diverse research fields from plant biology to immunology. The following chapters describe in-depth protocols and troubleshooting tips covering many of these applications.

Acknowledgments

The author is funded in part by the Pharmacogenetics Research Network (U01 GM63340), R21 CA113491 and R21 CA102461.

References

1. Rose, C. M., Marsh, S., Ameyaw, M. M., and McLeod, H. L. (2003) Pharmacogenetic analysis of clinically relevant genetic polymorphisms. *Methods Mol. Med.* **85,** 225–237.
2. Suarez, B. K., Taylor, C., Bertelsen, S., et al. (2005) An analysis of identical single-nucleotide polymorphisms genotyped by two different platforms. *BMC Genet.* **6,** S152.
3. Freimuth, R. R., Ameyaw, M. -M., Pritchard, S. C., Kwok, P. -Y., and McLeod, H. L. (2004) High-throughput genotyping methods for pharmacogenomic studies. *Current Pharmacogenomics* **2,** 21–33.
4. Ronaghi, M. (2001) Pyrosequencing sheds light on DNA sequencing. *Genome Res.* **11,** 3–11.
5. Ronaghi, M., Uhlen, M., and Nyren, P. (1998) A sequencing method based on real-time pyrophosphate. *Science* **281,** 363–365.
6. Marsh, S., King, C. R., Garsa, A. A., and McLeod, H. L. (2005) Pyrosequencing of clinically relevant polymorphisms. *Methods Mol. Biol.* **311,** 97–114.
7. Lanfear, D. E., Marsh, S., and McLeod, H. L. (2004) Caution with beta1-adrenergic receptor genotyping. *Clin. Pharmacol. Ther.* **76,** 185–186.
8. Aydin, A., Toliat, M. R., Bahring, S., Becker, C., and Nurnberg, P. (2006) New universal primers facilitate Pyrosequencing. *Electrophoresis* **27,** 394–397.
9. Pacey-Miller, T. and Henry, R. (2003) Single-nucleotide polymorphism detection in plants using a single-stranded pyrosequencing protocol with a universal biotinylated primer. *Anal. Biochem.* **317,** 166–170.
10. Pourmand, N., Elahi, E., Davis, R. W., and Ronaghi, M. (2002) Multiplex Pyrosequencing. *Nucleic Acids Res.* **30,** e31.
11. Wasson, J., Skolnick, G., Love-Gregory, L., and Permutt, M. A. (2002) Assessing allele frequencies of single nucleotide polymorphisms in DNA pools by pyrosequencing technology. *Biotechniques* **32,** 1144–1150.
12. Permutt, M. A., Wasson, J., Love-Gregory, L., et al. (2002) Searching for type 2 diabetes genes on chromosome 20. *Diabetes* **51,** S308–S315.
13. Skarke, C., Grosch, S., Geisslinger, G., and Lotsch, J. (2004) Single-step identification of all length polymorphisms in the UGT1A1 gene promoter. *Int. J. Clin. Pharmacol. Ther.* **42,** 133–138.
14. Jin, M., Gock, S. B., Jannetto, P. J., Jentzen, J. M., and Wong, S. H. (2005) Pharmacogenomics as molecular autopsy for forensic toxicology: genotyping cytochrome P450 3A4*1B and 3A5*3 for 25 fentanyl cases. *J. Anal. Toxicol.* **29,** 590–598.
15. Henningsson, A., Marsh, S., Loos, W. J., et al. (2005) Association of CYP2C8, CYP3A4, CYP3A5, and ABCB1 polymorphisms with the pharmacokinetics of paclitaxel. *Clin. Cancer Res.* **11,** 8097–8104.
16. Lee, S. S., Kim, K. M., Thi-Le, H., Yea, S. S., Cha, I. J., and Shin, J. G. (2005) Genetic polymorphism of CYP2C9 in a Vietnamese Kinh population. *Ther. Drug Monit.* **27,** 208–210.

17. de Jong, F. A., Marsh, S., Mathijssen, R. H., et al. (2004) ABCG2 pharmacogenetics: ethnic differences in allele frequency and assessment of influence on irinotecan disposition. *Clin. Cancer Res.* **10,** 5889–5894.
18. Aquilante, C. L., Lobmeyer, M. T., Langaee, T. Y., and Johnson, J. A. (2004) Comparison of cytochrome P450 2C9 genotyping methods and implications for the clinical laboratory. *Pharmacotherapy* **24,** 720–726.
19. Lanfear, D. E., Marsh, S., Cresci, S., Spertus, J. A., and McLeod, H. L. (2004) Frequency of compound genotypes associated with beta-blocker efficacy in congestive heart failure. *Pharmacogenomics* **5,** 553–558.
20. Eriksson, S., Berg, L. M., Wadelius, M., and Alderborn, A. (2002) Cytochrome p450 genotyping by multiplexed real-time dna sequencing with pyrosequencing technology. *Assay Drug Dev. Technol.* **1,** 49–59.
21. Ahluwalia, R., Freimuth, R., McLeod, H. L., and Marsh, S. (2003) Use of pyrosequencing to detect clinically relevant polymorphisms in dihydropyrimidine dehydrogenase. *Clin. Chem.* **49,** 1661–1664.
22. Garsa, A. A., McLeod, H. L., and Marsh, S. (2005) CYP3A4 and CYP3A5 genotyping by Pyrosequencing. *BMC Med. Genet.* **6,** 19.
23. Marsh, S., Xiao, M., Yu, J., et al. (2004) Pharmacogenomic assessment of carboxylesterases 1 and 2. *Genomics* **84,** 661–668.
24. Saeki, M., Saito, Y., Jinno, H., et al. (2003) Comprehensive UGT1A1 genotyping in a Japanese population by pyrosequencing. *Clin. Chem.* **49,** 1182–1185.
25. Fukunaga, A. K., Marsh, S., Murry, D. J., Hurley, T. D., and McLeod, H. L. (2004) Identification and analysis of single-nucleotide polymorphisms in the gemcitabine pharmacologic pathway. *Pharmacogenomics J.* **4,** 307–314.
26. King, C. R., Yu, J., Freimuth, R. R., McLeod, H. L., and Marsh, S. (2005) Interethnic variability of ERCC2 polymorphisms. *Pharmacogenomics J.* **5,** 54–59.
27. Rieder, M. J., Reiner, A. P., Gage, B. F., et al. (2005) Effect of VKORC1 haplotypes on transcriptional regulation and warfarin dose. *N. Engl. J. Med.* **352,** 2285–2293.
28. Innocenti, F., Undevia, S. D., Iyer, L., et al. (2004) Genetic variants in the UDP-glucuronosyltransferase 1A1 gene predict the risk of severe neutropenia of irinotecan. *J. Clin. Oncol.* **22,** 1382–1388.
29. Lotsch, J., Skarke, C., and Geisslinger, G. (2003) Simultaneous screening for three mutations in the ABCB1 gene. *Genomics* **82,** 503–510.
30. Pielberg, G., Day, A. E., Plastow, G. S., and Andersson, L. (2003) A sensitive method for detecting variation in copy numbers of duplicated genes. *Genome Res.* **13,** 2171–2177.
31. Soderback, E., Zackrisson, A. L., Lindblom, B., and Alderborn, A. (2005) Determination of CYP2D6 gene copy number by pyrosequencing. *Clin. Chem.* **51,** 522–531.
32. Wojnowski, L. and Brockmoller, J. (2004) Single nucleotide polymorphism characterization by mRNA expression imbalance assessment. *Pharmacogenetics* **14,** 267–269.
33. Teodoridis, J. M., Hall, J., Marsh, S., et al. (2005) CpG island methylation of DNA damage response genes in advanced ovarian cancer. *Cancer Res.* **65,** 8961–8967.

34. Tost, J., Dunker, J., and Gut, I. G. (2003) Analysis and quantification of multiple methylation variable positions in CpG islands by Pyrosequencing. *Biotechniques* **35,** 152–156.
35. Shaw, R. J., Liloglou, T., Rogers, S. N., et al. (2006) Promoter methylation of P16, RARbeta, E-cadherin, cyclin A1 and cytoglobin in oral cancer: quantitative evaluation using pyrosequencing. *Br. J. Cancer* **94,** 561–568.
36. Entz, P., Toliat, M. R., Hampe, J., et al. (2005) New strategies for efficient typing of HLA class-II loci DQB1 and DRB1 by using Pyrosequencing. *Tissue Antigens* **65,** 67–80.
37. Wu, H., Khanna, D., Park, G., et al. (2004) Interaction between RANKL and HLA-DRB1 genotypes may contribute to younger age at onset of seropositive rheumatoid arthritis in an inception cohort. *Arthritis Rheum.* **50,** 3093–3103.
38. Ringquist, S., Alexander, A. M., Styche, A., Pecoraro, C., Rudert, W. A., and Trucco, M. (2004) HLA class II DRB high resolution genotyping by pyrosequencing: comparison of group specific PCR and pyrosequencing primers. *Hum. Immunol.* **65,** 163–174.
39. Ramon, D., Braden, M., Adams, S., Marincola, F. M., and Wang, L. (2003) Pyrosequencing trade mark: A one-step method for high resolution HLA typing. *J. Transl. Med.* **1,** 9.
40. Hochberg, E. P., Miklos, D. B., Neuberg, D., et al. (2003) A novel rapid single nucleotide polymorphism (SNP)-based method for assessment of hematopoietic chimerism after allogeneic stem cell transplantation. *Blood* **101,** 363–369.
41. Ringquist, S., Alexander, A. M., Rudert, W. A., Styche, A., and Trucco, M. (2002) Pyrosequence-based typing of alleles of the HLA-DQB1 gene. *Biotechniques* **33,** 166–165.
42. Innings, A., Krabbe, M., Ullberg, M., and Herrmann, B. (2005) Identification of 43 Streptococcus species by pyrosequencing analysis of the rnpB gene. *J. Clin. Microbiol.* **43,** 5983–5991.
43. Pai, R., Limor, J., and Beall, B. (2005) Use of pyrosequencing to differentiate Streptococcus pneumoniae serotypes 6A and 6B. *J. Clin. Microbiol.* **43,** 4820–4822.
44. Sinclair, A., Arnold, C., and Woodford, N. (2003) Rapid detection and estimation by pyrosequencing of 23S rRNA genes with a single nucleotide polymorphism conferring linezolid resistance in Enterococci. *Antimicrob. Agents Chemother.* **47,** 3620–3622.
45. Rahim, A., Coutelle, C., and Harbottle, R. (2003) High-throughput Pyrosequencing of a phage display library for the identification of enriched target-specific peptides. *Biotechniques* **35,** 317–324.
46. Marsh, S., King, C. R., Garsa, A. A., and McLeod, M. L. (2005) Pyrosequencing of clinically relevant polymorphisms. In: *Pharmacogenomics: Methods and Protocols* (Innocenti, F., ed.), Humana Press, Totowa, NJ, pp. 97–114.

3

Web-Based Primer Design Software for Genome-Scale Genotyping by Pyrosequencing®

Steven Ringquist, Christopher Pecoraro, Ying Lu, Alexis Styche, William A. Rudert, Panayiotis V. Benos, and Massimo Trucco

Summary

Design of locus-specific primers for use during genetic analysis requires combining information from multiple sources and can be a time-consuming process when validating large numbers of assays. Data warehousing of genomic DNA sequences and genetic variations when coupled with software applications for optimizing the generation of locus-specific primers can increase the efficiency of assay development. Selection of oligonucleotide primers for PCR and Pyrosequencing® (SOP^3) software allows user-directed queries of warehoused data collected from the human and mouse genome sequencing projects. The software automates collection of DNA sequence flanking single-nucleotide polymorphisms (SNPs) as well as the incorporation of locus-associated functional information, such as whether the SNP occurs in an exon, intron, or untranslated region. SOP^3 software accepts three types of user-directed input consisting of gene locus symbols, SNP reference sequence numbers, or chromosomal physical location. For human polymorphisms, SOP^3 incorporates haplotype, ethnicity, and SNP validation attributes. The output is a list of oligonucleotide primers recommended for Pyrosequencing-based typing of genetic variations. SOP^3 is available at the Division of Immunogenetics computational server found at http://imgen.ccbb.pitt.edu.

Key Words: Bioinformatics; genetics; genomics; genotyping; sequencing.

1. Introduction

Computer software for designing primers for use during PCR are available from a number of sources ***(1–3)***. A principle use of these applications is to aid assay design when evaluating the association between phenotypical changes and the inheritance of genetic markers, such as single-nucleotide polymorphisms (SNPs) including nucleotide insertions and deletions. Commercial software is also available for generating PCR and sequencing primers and for use

From: *Methods in Molecular Biology, vol. 373: Pyrosequencing® Protocols*
Edited by: S. Marsh © Humana Press Inc., Totowa, NJ

during DNA sequencing. For example, the manufacturer of the Pyrosequencing® instrument Biotage, LLC provides assay design software aimed at developing primer sets suitable for use during Pyrosequencing. These applications provide access to PCR and sequencing primer design algorithms, but require user input of FASTA-formatted sequences along with identification of polymorphic residues that are to be evaluated. For genetic studies where many polymorphisms are being investigated, such as is necessary when investigating a set of candidate markers for association with a disease phenotype, this requirement creates a bottleneck during assay development. Improved approaches to assay design that provide integrated genomic sequence, genetic variant, and functional information increase the efficiency by which genotyping assays can be developed and validated by laboratory testing *(4)*.

SOP³ is a web-based software application that has been developed for use in Pyrosequencing assay design and is available at http://imgen.ccbb.pitt.edu *(5)*. The computational webserver warehouses an integrated biological database comprised of genetic variants, as well as DNA sequence of the human and mouse genomes, including validation of human SNPs within the HapMap comparative genome project. The software provides a batch analysis tool that accepts the names of as many as 100 genetic loci, 100 SNP identifiers, or a 100-kb section of genomic DNA under investigation. The application returns a set of primers for PCR and Pyrosequencing for all corresponding genetic variants listed in dbSNP along with flanking genomic DNA sequences contained within the human and mouse genome resources. Local warehousing of genomic sequences, making them accessible to the application, enables the software to provide the choice of locus-specific primer trios for PCR and Pyrosequencing for use in rapid genotype analysis.

2. Materials

2.1. System Configuration

1. The SOP³ application, genomic sequence, and genetic variation data from the human and mouse genome projects are warehoused on a computational webserver purchased from @Xi Computer (San Clemente, CA). The software application is written in preprocessor hypertext protocol v5.0.3 and is associated with a MySQL 4.1 database developed on a Linux SuSE Enterprise Server 8 for the AMD64 operating system with Apache web server v2.0.48. The computational server consists of dual AMD Opteron 246 64-bit processors with 1024 KB Cache, 8192 MB random access memory, and four 250 GB hard drives.

2.2. Connecting to the Webserver

1. The SOP³ website can be accessed over the internet at http://imgen.ccbb.pitt.edu. The application works best when viewed in Internet Explorer v6 or higher but can also be viewed in Firefox as well as Netscape. Internet browsers are available for

download from http://www.microsoft.com/windows/ie, http://www.mozilla.org, and http://browser.netscape.com, respectively.

3. Methods

The SOP^3 application has been designed for use with genotyping protocols and enables locus-specific PCR amplification of candidate SNPs for sequence analysis. Filtering of genetic variations for validated and HapMap-associated SNPs improves the efficiency of assay development for studying the role of DNA polymorphisms in determining the phenotype in models of genetic inheritance (**Table 1**). An advantage of the SOP^3 application is that design of genotyping studies can be developed using the assay design algorithm by filtering all available genetic variants for those specified by user-directed query, such as specifying human or mouse and using only validated nucleotide polymorphisms (*see* **Note 1**). The application accepts multiple queries of gene loci, SNPs, as well as chromosomal regions for analysis. SOP^3 provides multiple choices for selecting the most suitable genetic markers for evaluation during genotyping. The User Guide for operation of SOP^3 software is available online at http://imgen.ccbb.pitt.edu/sop3/userguide.

3.1. Basic Functions

1. Searching human or mouse genetic variations. The web interface to the SOP^3 application allows the user to specify whether to search for SNPs within the human or mouse genome. The default setting is for human genetic variants and is indicated on the website by the filled in button adjacent to the label for "Human." To specify searching of the mouse database the user should click the button next to "Mouse" provided on the website (**Fig. 1**).
2. Searching by locus name. To query the SOP^3 database for SNPs associated with up to 100 locus symbols the user can enter the symbol or a list of locus symbols into the textbox. Locus symbols can be typed directly into the textbox, entered using the copy and paste function on the user's computer, or by uploading from a text file. Queries are restricted to no more than 100 entries per batch. Use the check boxes provided to specify whether the query is for human or mouse genomes. Click the "Search" button to activate the application. Results will appear at the bottom of the page. Selecting the "Clear" button will refresh the application to the default settings.
3. Searching by SNP identifier. To query the SOP^3 database by SNP identifier the user can enter either a single reference sequence number or a list of polymorphism reference sequence numbers. Queries are restricted to no more than 100 entries per batch. Use the check boxes provided to specify whether the query is for human or mouse genomes. Click the "Search" button to activate the application. Results will appear at the bottom of the page. Selecting the "Clear" button will refresh the application to the default settings.
4. Searching by chromosomal location and range. To query the SOP^3 database by chromosomal range use the check boxes to specify whether the search will focus

Table 1
Classification of Human and Mouse Genetic Variants

Classification	Number of human SNPs	Number of mouse SNPs	Description
Validation status			
HapMap	35,126	NA	The genetic variant has been genotyped by the International HapMap Consortium and incorporated into the dbSNP resource.
Validated	384,831	464,137	Multiple submissions of the genetic variant has been reported within the dbSNP resource.
Function type			
Coding: syn. unknown	34	9	The genetic variant is within the coding region of a gene. The location of the polymorphism cannot be resolved because of an error in the alignment of the exon.
Intron	3,704,388	213,968	The variant is in the intron of a gene but not within two residues from the intron/exon boundary.
Nonsynonymous change	61,963	5478	The genetic variant is nonsynonymous for the codon within the gene. This class of allele is defined by a substitution and translation of the allele into the codon that results in a change to the amino acid specified by the exon reading frame.
No RNA_acc / Protein_acc	357,606	30,373	The polymorphism is within 2000 nucleotides 5' or 500 nucleotides 3' of a gene feature. The variant is not in the transcript for the gene.
Splice site	971	23	The variant is in the first two or last two nucleotides on the intron of a gene.
Synonymous change	48,702	9548	The genetic variant is synonymous for the codon within the gene. This class of allele is defined by a substitution and translation of the allele into the codon that results in no change to the amino acid specified by the exon reading frame.
Untranslated region	668,282	43,414	The nucleotide polymorphism is in the transcript of a gene but not in the coding region of the mRNA.

Geographic population			
Cent. Asia	46	NA	Samples from Russia and its satellite republics and from nations bordering the Indian Ocean between East Asia and the Persian Gulf regions.
Cent/S. Africa	30	NA	Samples from nations south of the Equator, Madagascar, and neighboring island nations.
Cent/S. America	11	NA	Samples from Mainland Central and South America and island nations of the western Atlantic, Gulf of Mexico, and Eastern Pacific.
E. Asia	7447	NA	Samples from eastern and south eastern Mainland Asia and from Northern Pacific island nations.
Europe	535	NA	Samples from Europe north and west of Caucasus Mountains, Scandinavia, and Atlantic islands.
Multinations	13,004	NA	Samples that were designated to maximize measures of heterogeneity or sample human diversity in a global fashion.
N. America	11,438	NA	All samples north of the Tropic of Cancer, including defined samples of US caucasians, African Americans, Hispanic Americans, and the National Human Genome Research Initiative (NHGRI) poly morphism discovery resource.
N.E. Africa and Mid. East	27	NA	Samples collected from North Africa (including the Sahara desert), East Africa (south to the Equator), Levant, and the Persian Gulf.
Pacific	144	NA	Samples from Australia, New Zealand, Central and Southern Pacific Islands, and Southeast Asian peninsular/island nations.
Unknown	479	NA	Samples with unknown geographic provinces that are not global in nature.
W. Africa	174	NA	Sub-Saharan nations bordering the Atlantic north of the Congo River and central/southern Atlantic nations.

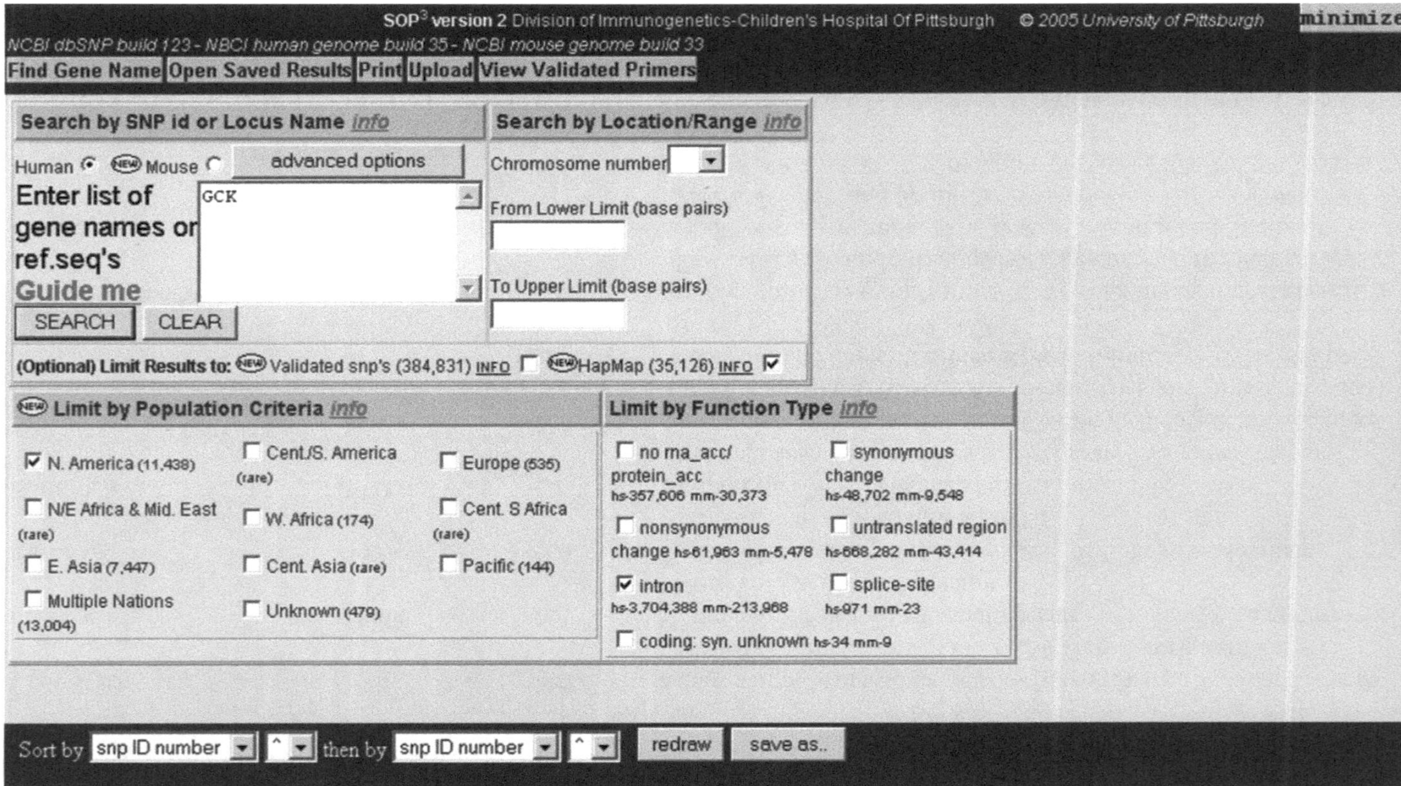

Fig. 1. User interface to the SOP[3] web-based software application. The example illustrates a search for polymorphisms associated with the human glucokinase locus, *GCK*. The query is limited to single-nucleotide polymorphisms linked to the HapMap project, filtered for genetic variants found in populations within North America, and for polymorphisms associated with intron elements of the RNA transcript.

Table 2
Maximum Allowable Upper Limit for Screening Chromosome Range

Human chromosome	Maximum range	Mouse chromosome	Maximum range
1	245,442,500	1	195,198,653
2	242,818,021	2	181,685,801
3	199,450,386	3	160,571,871
4	191,400,945	4	154,132,574
5	180,837,593	5	149,217,787
6	170,972,421	6	149,553,910
7	206,556,958	7	133,031,105
8	146,272,185	8	128,674,990
9	138,428,984	9	124,140,960
10	135,412,916	10	130,567,357
11	134,451,003	11	121,607,694
12	132,389,146	12	114,933,529
13	114,127,336	13	116,456,691
14	106,360,250	14	117,011,917
15	100,337,960	15	104,102,711
16	88,821,548	16	98,800,952
17	80,652,345	17	93,538,550
18	76,116,152	18	90,878,013
19	63,806,020	19	60,667,351
20	62,435,629	X	160,070,598
21	46,943,948	Y	47,759,179
22	49,534,318		
X	154,823,225		
Y	57,700,652		

Values for maximum range were determined using the highest chromosomal position number associated with a SNP on the human or mouse genomes using build 35 or 33, respectively.

on human or mouse genomes. Next, use the "Chromosome Number" dropdown menu to specify which chromosome is to be used. Enter lower and upper limits for the nucleotide positions that should be screened. Queries are restricted to regions no greater than 100,000 nucleotides. The maximum allowable upper limit for screening a range of nucleotides on each chromosome corresponds to the length of the chromosome and is indicated in **Table 2**. Click the "Search" button to activate the application. Results will appear at the bottom of the page. Selecting the "Clear" button will refresh the application to the default settings.

5. Filtering the results to show only validated SNPs. Selection of the "Validated SNPs" search filter directs the application to screen genetic polymorphisms for inclusion in the list of validated polymorphisms provided by dbSNP. There are currently 384,831 human and 464,137 mouse polymorphisms associated with this attribute. Validation status is defined as a genetic variant that has been reported

by multiple submissions, is linked to frequency data for the SNP in a population, or was submitter validated by an updated submission from the original report.

6. Filtering the results to show only HapMap-genotyped SNPs. Selection of the "HapMap" search filter directs the application to screen genetic polymorphisms for those validated polymorphisms that are also included in the International HapMap Consortium haplotyping mapping project incorporated into the dbSNP database. There are currently 35,126 human polymorphisms associated with this attribute.
7. Filtering the results by limiting the population criteria. Selection of the "Limit By Population Criteria" filter directs the application to screen genetic polymorphisms by geographic origin of the sample. Selection of more than one population filter will return those polymorphisms that meet either of the criteria (i.e., a Boolean OR operation). The number of polymorphisms associated with each population group is indicated on the application's homepage.
8. Limiting the search by function type. Genetic variants may be filtered by whether they are near a gene (locus region), in an untranslated region (UTR), in an intron (intron), or in a splice site (splice site). If the variation is in a coding region, then the functional class of the variation depends on how each allele may affect the translated peptide sequence, e.g., nonsynonymous change or synonymous change. Selection of more than one of the function-type filters will return those polymorphisms that meet either of the criteria (i.e., a Boolean OR operator). A definition of these filters as well as the number of polymorphisms associated with each function type, for human and mouse, is indicated in **Table 1**.
9. Initiating design of primers for PCR and Pyrosequencing. Selecting the "Search" button will activate the application. Locus-specific primers for PCR and Pyrosequencing will be returned for those SNPs that meet the filtering criteria selected in the user input section of the web interface.
10. Reporting the results. After submitting a search the application will present the results at the bottom frame of the website. Each report consists of four sections, attribute bar, candidate primer bar, DNA sequence flanking the polymorphism, and computer-simulated Pyrosequencing data for heterozygous and each homozygous genotype (**Fig. 2**).
11. The locus attribute bar provides information linked to the polymorphism consisting of reference sequence number, identity of the polymorphism, locus symbol, physical location on the chromosome, heterozygosity value, and function attribute. If the filters for "Validated SNPs," "HapMap," or "Limit By Population Criteria" were selected that information will also appear in the attribute bar.
12. The candidate primer bar contains the oligonucleotide sequence (written 5' to 3') for the candidate PCR forward and reverse primers as well as the Pyrosequencing primer (*see* **Note 2**). The nomenclature used for designating a forward PCR primer is that it is on the same side of the SNP as the Pyrosequencing primer. The nomenclature for designating a reverse PCR primer is that it is on the opposite side of the SNP as the Pyrosequencing primer. The reverse PCR primer should be synthesized with a 5' biotin-tetraethylene glycol modification during the Pyrosequencing protocol.

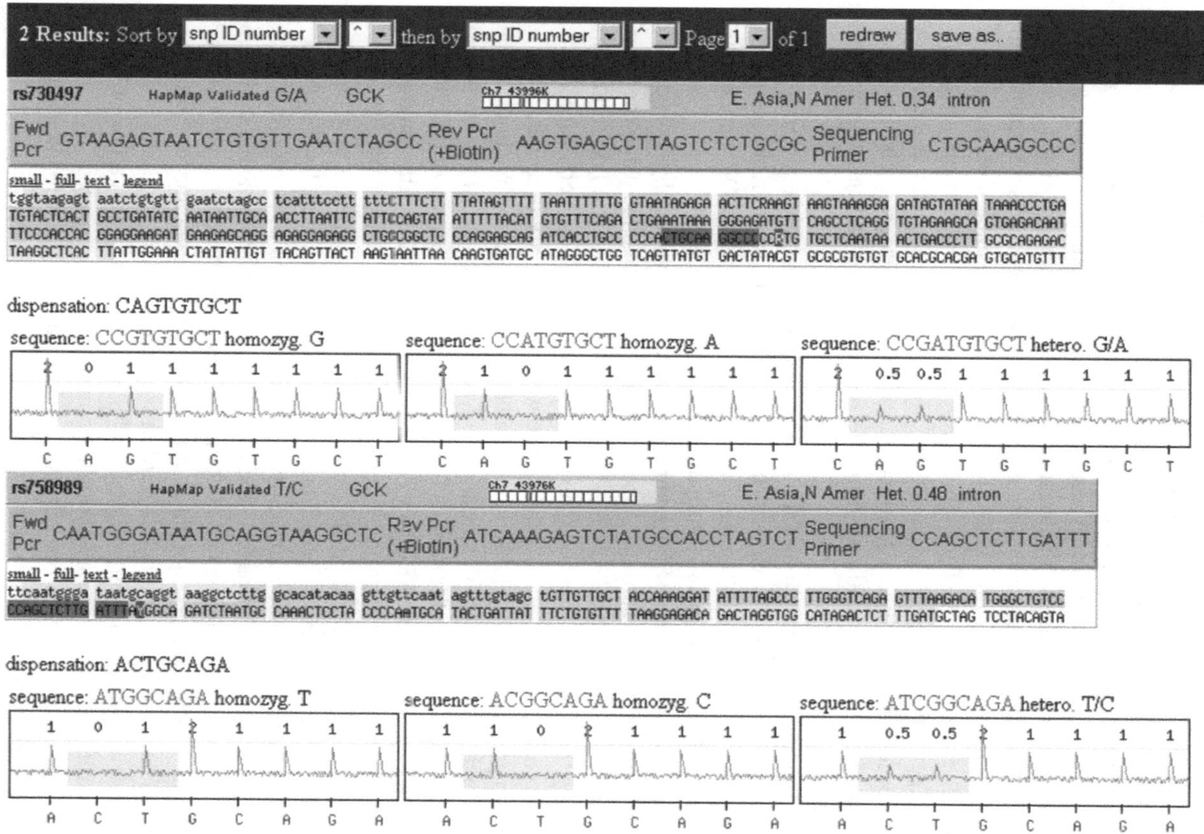

Fig. 2. Results of the query for single-nucleotide polymorphisms (SNPs) linked to the human glucokinase locus, *GCK*. There are two entries in SOP[3] v2 corresponding to the query. The attribute bar, candidate primer bar, DNA sequence, and simulated Pyrosequencing® data are shown. The DNA sequence box indicates the location of the PCR primers, Pyrosequencing primer, region of amplified DNA, and location of the SNP.

13. The DNA sequence flanking the SNP is indicated in the third panel of the results frame. The user may view the "Full" sequence that was evaluated, consisting of the polymorphism and 1000 flanking residues. The DNA sequence may also be viewed in FASTA format by selecting the "Text" button. The option for viewing the flanking sequence as "Small" will display only the sequence amplified by the locus-specific PCR primers (*see* **Note 3**). The color code in the bottom frame shows PCR primers indicated in yellow, Pyrosequencing primer in red, the SNP is colored blue, the amplified region in orange, and nearby polymorphisms shown in gray. Lowercase characters indicate the location of repeat masker-identified genomic regions. A "Legend" button providing a key to the color coding is also provided.
14. Simulated Pyrosequencing data is displayed in the bottom panel of each result. A simulated Pyrogram is drawn for each genotype, i.e., heterozygous and both homozygous genotypes. A nucleotide dispensation order for evaluating these data in the Pyrosequencing software is illustrated above the Pyrogram charts. Informative dispensations for distinguishing the genotypes are indicated in yellow (*see* **Note 4**).
15. Minimization and maximization of the input/output frames can be accomplished by clicking the "Minimize/Maximize" button located in the upper right-hand corner of the web interface. A full-screen view of the results of the SOP^3 application is available, allowing easier browsing of the results frame.
16. Sorting the results by reference SNP number, heterozygosity value, physical position on the chromosome, gene name, or function can be accomplished using the "Sort By" dropdown menus. The "Sort By" function is located in the horizontal bar found along the middle of the web interface. Results may be viewed in ascending "^" or descending "v" order. Results on the web interface are presented 20 at a time. When greater than 20 results are returned they can be viewed by choosing the "Page" dropdown menu. The total number of results and pages of results is also indicated. After selection of functions for sorting and page display the results can be refreshed by clicking the "Redraw" button located on the middle toolbar.
17. Results from the SOP^3 application can be saved in an Excel formatted text file. Select this function by clicking the "Save As." This will open a text box for designating a file name and a "Save" button for uploading the results to the SOP^3 server. The saved file can be viewed by selecting the "Open Saved Results" button located at the top of the page. Saved files are listed alphabetically and each file contains a tab-delimited list of PCR and sequencing primers, as well as SNP attributes such as reference sequence number, heterozygosity value, and function attribute. Selection of the "Save All" button will allow all the results to be transferred to the web server. Unselecting of the "Save All" button will allow only the page of results presently displayed in the bottom frame to be saved.
18. Selecting the "Clear" button located in the upper frame of the web interface will refresh the application to the default settings.

3.2. Advanced Options for PCR Primer Design

1. Selection of the "Advanced Options" button opens a list of PCR and sequencing primer design settings used in the assay design algorithm. This function

allows the user to specify customized settings for primer selection that are different from the default settings. Clicking the "Advanced Options" a second time will hide these values. Selecting the "Clear" button will refresh the application to the default settings.

2. The display for PCR primer design settings allows the user to adjust the primer selection criteria.
3. "Melting Temperature (T_m) °C" default setting is 60°C. T_m is calculated using the relationship (T_m = 16.6*log[cation concentration] + 41*[fraction of GC] + 81.5, where the concentration of cation was estimated to be 0.1 *M*) described by Schildkraut and Lifson ***(6)***. User-directed changes to this value instruct the application to design primers to the indicated minimum T_m value.
4. The "Minimum Foldover" filter is set at a default of five. This value reflects the minimum allowable number of contiguous base-paired residues that can occur within a candidate PCR primer. The function allows user-directed input in order to minimize the formation of potential primer secondary structure that can interfere with PCR efficiency.
5. The "Unique N'mer Maximum Length" filter default setting is six and specifies the maximum number of 3'-end residues that meet the requirement of only occurring once within 1000 nucleotides flanking either side of the polymorphism.
6. The "A/T Test" filter specifies that at least one A or T residue will occur in the last three 3' nucleotides of the PCR primer.
7. The "Residue Thresholds" filter specifies the frequency range for each A, T, C, G nucleotide residue in the candidate PCR primer. The default setting for this filter is minimum 14% and maximum 40% nucleotide composition.
8. The "Flank Length (Maximum 1000)" setting default setting is 1000 and denotes the length of DNA flanking the polymorphism that is used to screen for suitable PCR primers.
9. The filter "Use Repeat Masker" indicates whether the results of RepeatMasker (repeatmasker.org) will be used when evaluating the DNA sequence flanking the SNP. When selected, residues that are identified as being included in a repetitive region will be masked off and will not be considered when making the PCR primer, thus generating primers that will not anneal to regions of repetitive DNA. Using this option will generally decrease the number of primers found, but will better ensure that primers do not anneal to multiple sites.
10. The filter for "PCR Product Size" determines the range of length of an acceptable amplified PCR product. The default setting for this filter is minimum 200 bp and maximum 500 bp.

3.3. Advanced Options for Pyrosequencing Primer Design

1. User-directed changes to the Pyrosequencing primer design settings are accomplished using the three interactive boxes available on the lower panel of the "Advanced Options" window.
2. The "Melting Temperature (T_m) °C" default setting is 40°C. T_m is calculated using the relationship (T_m = 16.6*log[cation concentration] + 41*[fraction of GC] + 81.5, where the concentration of cation was estimated to be 0.1 *M*) described

by Schildkraut and Lifson *(6)*. User-directed changes to this value instruct the application to design each sequencing primer to the indicated minimum T_m value.

3. The filter for "T+A Percentage" evaluates the design of the candidate sequencing primer for % AT content ensuring that this value falls within the specified range. The default range is minimum 30% and maximum 65%.
4. The interactive box for designating the "Distance From SNP" provides a tool for specifying the minimum number of bases away from the SNP to initiate the design of the Pyrosequencing primer. The default range is minimum 1 and maximum 30 nucleotides. The software algorithm is designed to choose the closest Pyrosequencing primer to the SNP but with the range of distance designated by the "Distance From SNP" function.

3.4. Exploring the Website

1. General information regarding warehoused genomic and genetic variant data is indicated in the masthead of the web interface. For example, the build of dbSNP and human and mouse DNA sequences used for v2 of SOP^3 are build 123, 35, and 33, respectively. Updates to the SOP^3 database are performed twice a year and are indicated in the masthead.
2. The "Find Gene Name" button opens a link to the search page for determining the locus symbol for either human or mouse genes as specified by the user. The "Fine Gene Name" search box allows simple Boolean searches. It allows and, or, not. It uses "*" to indicate a wildcard character and quotation marks to designate a phrase. Queries are submitted using the "Submit Query" button.
3. The "Open Saved Results" button opens a link to the list of saved files available on the SOP^3 web server. Selection of a file provides a tab-delimited document consisting of the locus symbol, reference sequence number, allele, heterzygosity value, chromosome location, primer sequences, and size of expected PCR product. The tab-delimited document can be saved to the user's computer and opened in Microsoft Excel.
4. The "Print" button provides a link to the user's print command for use in obtaining a hardcopy of the results.
5. The "Upload" button opens a "Text Box," "Browse" button, "Submit" button, and "Cancel" button for uploading to the web server a text file containing a batch of locus symbols or reference sequence numbers from a file located on the user's computer. Selection of the "Browse" button prompts the user to indicate the location of a carriage return delimited text file. Selection of the "Submit" button uploads the query batch to "Search By SNP ID Or Locus Name," i.e., the application's main entry box. The application is initiated by selecting the "Search" button. Selection of the "Cancel" button removes the upload feature from the web interface and allows the user to query genomic SNPs using the basic functions available on the website.
6. The "View Validated Primers" button opens a link to the primer trios that have been made available for general use. They are listed in a tabulated format and include reference sequence, locus, forward PCR primer, reverse PCR primer, Pyrosequencing primer, and PCR product length.

3.5. Online User Guide

1. The "Guide Me" button opens a link to the online user guide for the SOP³ application. The user guide is updated along with changes to the application and the warehoused database (*see* **Note 5**). Explanations of the choices available when using the application are indicated in alphabetical order.
2. The "Info" buttons open information links for selected functions available on the web interface. For example, selection of the "Info" button next to the "Limit By Function Type" heading opens a new browser window to the online user guide describing the use of this filter when selecting SNPs for genetic analysis.

4. Notes

1. Software applications for primer design are available from several sources. Commonly used web-based applications for designing PCR primers are available at the Primer3 (http://_frodo.wi.mit.edu) and IDT Corporation (http://www.idtdna.com) websites *(1)*. Both applications generate oligonucleotide sequences for locus-specific PCR amplification. They are limited, however, in that they do not provide a sequencing primer for Pyrosequencing. Another application, specific for Pyrosequencing, is Assay Design Software available commercially from Biotage, LLC. This application provides access to a primer design algorithm for locus-specific PCR and SNP-specific Pyrosequencing. The application, however, requires user input of each SNP and DNA flanking sequence. The Biotage Assay Design Software can be operated using a batch mode once the sequences are arranged in a text file on the user's computer. These applications and the SOP³ software provide complimentary solutions for selecting robust primers for use during pyrosequencing-based typing of genetic variants.
2. Addition of a unique nucleotide motif to the 5'-end of the forward PCR primer occurs in order to avoid formation of secondary structure in the biotinylated template strand, which can lead to competing sequencing signal owing to self priming during Pyrosequencing *(7–9)*. The nucleotide motif is chosen from a list of all possible nucleotide combinations that do not occur within the DNA sequence flanking the SNP.
3. Validating PCR primers designed by the computer application should be performed by PCR amplification followed by analysis of the product with agarose gel electrophoresis. Selection of primers assumes a standard PCR condition using an annealing temperature of 60°C. It is recommended that, when possible, PCR product yield be examined at 1 and 2 m*M* $MgCl_2$ and over an annealing temperature range from 54 to 65°C using a gradient thermal cycler instrument. PCR primers are considered validated if they result in a single amplified DNA product of the expected size as visualized by agarose gel electrophoresis.
4. Validating Pyrosequencing primers is accomplished by performing the Pyrosequencing reaction using a control DNA of known quality and combined with a set of negative control Pyrosequencing reactions. The negative control reactions are designed to allow troubleshooting of the sequencing reaction by analyzing the level of background signal associated with the sequencing primer, biotinylated reverse PCR primer, and possible alternative PCR products. Recom-

mended negative control reactions are as follows: (1) PCR product without the addition of Pyrosequencing primer; (2) PCR negative control reaction with Pyrosequencing primer; (3) PCR negative control reaction without Pyrosequencing primer; (4) Biotinylated reverse PCR primer alone; (5) Pyrosequencing primer alone; (6) Biotinylated PCR primer with Pyrosequencing primer. Failure to positively genotype samples using SOP^3 designed primers for PCR and Pyrosequencing can often be the result of background signal associated with the negative control reactions. Isolation of the background signal can be helpful in redesigning assays for selection of new primer trio sets.

5. Contact information for web server administration. The SOP^3 application is maintained by the Division of Immunogenetics at the Children's Hospital of Pittsburgh. Inquiries concerning customized searches of the database or reports of errors in the application should be addressed to Steven Ringquist (email: smr73@pitt.edu).

Acknowledgments

This work was supported by grants from the NIH, U19-AI056374-01 Autoimmunity Centers of Excellence (SR and MT), RO1DK24021(MT) from the National Institutes of Health, and ERHS no. 00021010 (SR and MT) from the Department of Defense. PVB was supported partially by the NSF grant MCB0316255.

References

1. Rozen, S. and Skaletsky, H. (2000) Primer3 on the WWW for general users and for biologist programmers. In: *Bioinformatics Methods and Protocols,* (Krawetz, S. and Misener S., eds.), Humana Press, Totowa, NJ, pp. 365–386.
2. Varotto, C., Richly, E., Salamini, F., and Leister, D. (2001) GST-PRIME: a genome-wide primer design software for the generation of gene sequence tags. *Nucl. Acids Res.* **29,** 4373–4377.
3. Emrich, S. J., Lowe, M., and Delcher, A. L. (2003) PROBEmer: a web-based software tool for selection optimal DNA oligos. *Nucl. Acids Res.* **31,** 3746–3750.
4. Stein, L. D. (2003) Integrating biological databases. *Nat. Rev. Genet.* **4,** 337–345.
5. Ringquist, S., Pecoraro, C., Gilchrist, M. S., et al. (2005) SOP^3v2: web-based selection of oligonucleotide primer trios for genotyping of human and mouse polymorphisms. *Nucl. Acids Res.* **33,** W548–W552.
6. Schildkraut, C. and Lifson, S. (1965) Dependence of the melting temperature of DNA on salt concentration. *Biopolymers* **3,** 195–208.
7. Ronaghi, M., Pettersson, B., Uhlen, M., and Nyren, P. (1998) PCR-introduced loop structure as primer in DNA sequencing. *Biotechniques* **25,** 876–878.
8. Ringquist, S., Alexander, A. M., Rudert, W. A., Styche, A., and Trucco, M. (2002) Pyrosequence based typing of alleles of the HLA-DQB1 gene. *Biotechniques* **33,** 166–175.
9. Utting, M. J., Hampe, J., Platzer, M., and Huse, K. (2004) Locking of 3' ends of single-stranded DNA templates for improved Pyrosequencing performance. *BioTechniques* **37,** 66–73.

4

Pyrosequencing®

A Simple Method for Accurate Genotyping

Cristi R. King and Tiffany Scott-Horton

Summary

Pharmacogenetic research benefits first-hand from the abundance of information provided by the completion of the Human Genome Project. With such a tremendous amount of data available comes an explosion of genotyping methods. Pyrosequencing® is one of the most thorough yet simple methods to date used to analyze polymorphisms. It also has the ability to identify tri-allelic, indels, short-repeat polymorphisms, along with determining allele percentages for methylation or pooled sample assessment. In addition, there is a standardized control sequence that provides internal quality control. This method has led to rapid and efficient single-nucleotide polymorphism evaluation including many clinically relevant polymorphisms. The technique and methodology of Pyrosequencing is explained in this chapter.

Key Words: Pyrosequencing®; genotype; single-nucleotide polymorphism; SNP; pharmacogenetics; pharmacogenomics.

1. Introduction

Six years after the Human Genome Project was completed, scientists are still amazed at the vast amount of genetic information that can be derived from the project. Specifically, pharmacogenetic research benefits first-hand from this landfill of information, in which each day proves to be more insightful. The premature estimation of 1.42 million single-nucleotide polymorphisms (SNPs) predicted as being present in the human genome seems quite primitive when compared with the current 125 build of the public SNP repository, dbSNP (http://www.ncbi.nlm.nih.gov/SNP/index.html) *(1)*. Today, the repository contains 27,189,291 submissions and almost 5 million validated SNPs, and it continues to grow at a rate of about 90 SNPs per month.

From: *Methods in Molecular Biology, vol. 373: Pyrosequencing® Protocols*
Edited by: S. Marsh © Humana Press Inc., Totowa, NJ

The distinctions found between individuals in disease susceptibility and drug metabolism are largely owing to polymorphisms located in the exonic, untranslated, and intronic regions of genes, ultimately producing alterations in protein expression, function, and behavior. As the number of reported SNPs increase, the ability to detect and evaluate them improves. The primary goal of pharmacogenetics is to determine how genetic inheritance influences response to drugs, which will lay the foundation for selecting the most advantageous drug therapy and dosage for individual patients.

Pyrosequencing® is one of the most thorough yet simple methods to date used to analyze SNPs. This method involves the utilization of ATP to produce light that generates peaks in the Pyrogram®. Although there is no restriction enzyme site involved in the process, the PCR and the internal primer sites are not based on a standard size or position. However, Pyrosequencing does allow a standardized control sequence for each sample that serves as an internal quality control that checks and fails samples that are not in concurrence with the standard. This method is not limited to the detection of just one allele. It also has the ability to identify tri-allelic, indels, and short-repeat polymorphisms, as well as determining allele percentages for methylation or pooled-sample assessment. More than 2500 individual genotypes can be assessed daily on a 96-well platform. This method has led to rapid and efficient SNP evaluation including many clinically relevant polymorphisms *(2–6)*.

2. Materials

2.1. DNA Template

Pyrosequencing assays can be used with DNA from cell lines, blood, serum, plasma, paraffin-embedded tissue, frozen tissue, and whole-genome amplified products (*see* **Note 1**). cDNA may also be used for Pyrosequencing.

2.2. PCR

1. Free or custom bought primer design software.
2. 1–5 ng DNA template (*see* **Note 2**).
3. PCR Mastermix (*see* **Note 3**).
4. *Taq* polymerase (*see* **Note 4**).
5. DNase- and RNase-free 18.2 MΩ-cm water.
6. DNA oligonucleotide primers.
7. Unskirted 96-well PCR trays.
8. 96-Well thermal adhesive sealing film or silicon mat for covering 96-well plates in a thermocycler.
9. Thermocycler (with 96-well capacity with a heated lid option).

2.3. Gel Electrophoresis

1. Agarose.

2. 50X TAE buffer: to make 1 L add 242 g Tris-base, 57.1 mL glacial acetic acid, and 18.6 g EDTA to 18.2 MΩ water. Store at room temperature and dilute to 1X (with water) when ready to use. Premade 50X TAE is also available for diluting.
3. Microwave.
4. Ethidium bromide: add 4 µL of 10 mg/mL ethidium bromide/100 mL agarose gel after heating in the microwave. Ethidium bromide is highly carcinogenic so make sure gloves are worn when handling. In addition, proper disposal and handling of spills should be known prior to use.
5. 6X Loading dye: for a 100-mL solution add 30 mL glycerol, 70 mL water, and a pinch of bromophenol blue and xylene cyanol FF (to desired color). Store at room temperature.
6. Gel apparatus that includes the casting tray, gel tank, and power supply.
7. Ultraviolet (UV) gel documentation system with thermal printer.

2.4. Peforming PCR for Pyrosequencing

1. Centrifuge with rotor/buckets to handle 96-well plates.
2. 2X Binding buffer: for 1 L add 1.21 g Tris, 117 g NaCl, and 0.292 g EDTA to 18.2 MΩ water. Adjust pH to 7.6 with 1 *M* HCl, sterile filter, then add 1 mL Tween-20.
3. Sepharose bead mix: for one 96-well plate combine 240 µL streptavidin-coated sepharose beads, 4560 µL 2X binding buffer, and 3600 µL 18.2 MΩ water. Excess sepharose bead mix may be stored in a glass bottle at 4°C. The magnetic bead-processing protocol for a PSQ96 or PSQ96MA is described elsewhere *(7)*.
4. 96-Well plate shaker, e.g., Eppendorf thermomixer (Fisher Scientific, Hampton NH).
5. Vacuum prep tool and troughs (Biotage, Westborough, MA) (*see* **Note 5**).
6. 70% Ethanol in 18.2 MΩ-cm water.
7. 0.2 *M* NaOH in 18.2 MΩ-cm water.
8. Washing buffer: for 1 L add 1.21 g Tris to 18.2 MΩ-cm water, adjust pH to 7.6 with 4 *M* acetic acid, and sterile filter.
9. Annealing buffer: for 1 L add 2.42 g Tris and 0.43 g magnesium acetate-tetrahydrate to 18.2 MΩ water, adjust pH to 7.6 with 4 *M* acetic acid, and sterile filter.
10. Pyrosequencing primer (Internal primer).
11. A 96-well Pyrosequencing plate.
12. Heat block (capable of reaching at least 80°C).
13. Pyrosequencing plate adaptor base and iron set (Biotage).
14. Adhesive sealing film for the 96-well Pyrosequencing plates.

2.5. Pyrosequencing

1. PSQhs96 or PSQhs96A Pyrosequencing system with Pyrosequencing 96A v1.1 or 96MA software or higher. A detailed protocol for the PSQ96 or PSQ96MA has been described previously *(7)*.
2. PSQ cartridge, capillary-dispensing tips (or nucleotide-dispensing tips), and reagent dispensing tips for hsPSQ96 and hsPSQ96A (Biotage).

3. Pyrosequencing hs reagent kit (Biotage).
4. DNase- and RNase-free 18.2 MΩ-cm water.
5. Microcentrifuge.
6. 1X TE: 10 m*M* Tris-HC1, m*M* EDTA (pH 8.0).

3. Methods

Pyrosequencing consists of sequencing by synthesis, which is a four-enzyme process performed in a single well. Nucleotides are incorporated to the open 3' DNA strand in which pyrophosphate is released and used in a sulfurylase reaction emitting ATP. The ATP is then used by luciferase, which is converted to oxyluciferin. Light is discharged as a result of the reaction and collected by a CCD camera. The light is assembled into a readable format and represented as peaks, commonly called Pyrogram charts. However, there are some cases where the nucleotide is not incorporated in the reaction, therefore, pyrophosphate is not released and the unutilized nucleotide is then removed by apyrase causing its degradation.

3.1. PCR Primer Design

1. There is a multitude of PCR design software available, some of which are available on the Internet for free, and others that are customized for a fee. The software allows one to find the gene of interest and locate both known or unknown polymorphisms that may exist. Polymorphisms such as SNPs, indels, repeats, and others may be in various positions in the gene, in which the design specifications are the same. While designing primers, ensure that there are no SNPs before or after the actual polymorphism of interest, as well as no overlapping of SNPs. If this occurs, it will be very challenging to design an optimal set of primers. In addition, if the sequence does not seem to exist as it is in the genome, try reversing the sequence manually.
2. Primer length and amplicon: the length of a primer should be between 15 and 30 bases long. The optimum length is 20 bases per primer with the GC:AT ratio at 50%. However, this is all dependent on the location of the polymorphism. In the most favorable conditions, the amplicon should extend from 100 to 200 bases, even for fragmented DNA. However, there are some cases where the amplicon extends 300–500 base pair, which may offer a suitable environment for template sources as well (*see* **Note 6**).
3. Melting temperature (T_m): the optimal melting temperature for the forward and reverse primer is 60°C. In fact, the forward and reverse primers should always be within two degrees of each other if they are not the same temperature. The T_m may range from 50 to 69°C in order to include those primers that are not located in the best position. In addition, A/T bases tend to lower the T_m, whereas C/G bases increase the T_m. The primer sequence may have to be adjusted either by adding to shortening the sequence in order to obtain the best T_m for a primer.

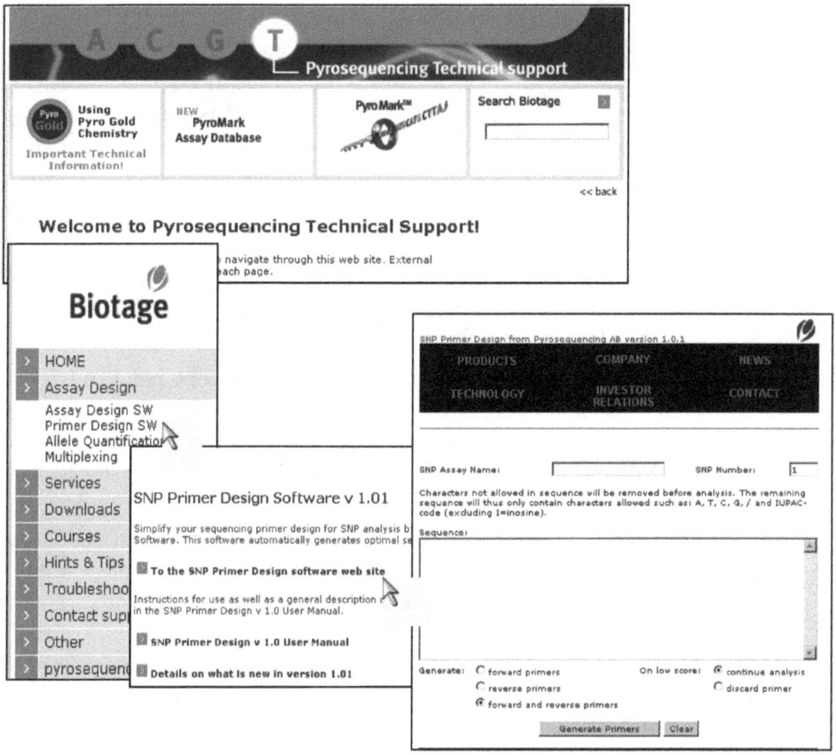

Fig. 1. Pyrosequencing® technical support website for Pyrosequencing primer design.

4. Primer specificity: the NCBI Blast program (http://www.ncbi.nlm.nih.gov/blast/) offers an incredible screening mechanism for primers to be checked across the human genome sequence. Designing primers for gene family members or genes with known pseudogenes can be complicated. As a result, cross-hybridization of primers may cause background noise in the sequence, false-positive results, and/or reduced or no signal. One of these primers must be biotinylated at the 5'-end (dependent on the orientation of the Pyrosequencing primer).

3.2. Pyrosequencing Primer Design

1. Pyrosequencing technical support: http://techsupport.Pyrosequencing.com is the site that offers complimentary services to all registered users to generate the most favorable Pyrosequencing primers (internal primers). An email address and password are required for all users. The entire amplicon sequence, which includes the forward and reverse sequences, should be cut and pasted into the SNP assay design software (**Fig. 1**). The SNP is recognized by the Universal Code and the primer that is generated extends from the 3'-end.

2. The software: there should be a default setting made to search for both the forward and reverse primers. This allows for an increased chance to acquire an optimum primer sequence, unless multiplexing is required (*see* **Note 7**). The software lists its results based on score. One hundred is the top score and descends as the primer quality degrades. Thus, a score that ranges from 90 to 100 is considered "high," whereas lower scores are considered either "medium" or "low." However, just because a score is not ranked highly does not mean the primer is not useable. The software deems unusable primers as those detected with an asterisk by the score (*see* **Note 8**).
3. Critical scoring parameters: the software provides parameters that will likely raise, lower, or flag (asterisk) a Pyrosequencing primer.
   a. Template loops: the more loops present indicate a greater chance of background noise found in the sequence, but should not affect the overall score. However, if there are too many loops the software will flag the primer (*see* **Note 9**), particularly in areas that have a high GC concentration. Also, loops that contain less than four bases should be avoided to minimize background noise.
   b. Mispriming: occurs when the internal primer has the ability to anneal to multiple positions within the amplicon. The annealed regions of the 3'-ends begin to integrate nucleotides that produce inaccurate genotypes or intense background noise.
   c. Duplex formation: occurs when the internal primer dimerizes with itself. Thus, the proper annealing does not take place, which may also result in background noise or a reduced signal.
   d. Hairpin loop: these loops occur when a primer forms secondary structures in which there is not enough primer left for the reaction, and the signal may be reduced or nonexistent as well.
4. Noncritical scoring parameters: there are also some parameters on the software that are not as critical but may lower the score.
   a. Repeated bases at SNP sequence: repeated bases surrounding the SNP may reduce the score of a Pyrosequencing primer, which may be unavoidable in some cases because of the stabilized position of the SNP. Generally, there is a three repeat base pair limit, which the software can accommodate without a problem. A repeat of four or more base pairs will cause difficulty in discerning peak heights in the Pyrogram.
   b. Primer length: although the length of a primer may not reduce its reaction, primers with more than 15 bp tend to have lower scores.

3.3. PCR Gradients for Optimization

1. New assays require primer optimization of magnesium concentration and temperature. The use of a thermocycler with a gradient block allows for PCR with different magnesium concentrations to be analyzed as in **Subheading 1.1.** Temperature optimization only is required for the premade PCR mix. An example of a PCR gradient is as follows:

a. Mastermix: 130 μL Amplitaq Gold PCR mastermix (Applied Biosystems, Foster City, CA), forward primer (10 p*M* final concentration), reverse primer (10 p*M* final concentration), and 13 μL DNA up to 260 μL with 18.2 MΩ-cm water. Add 20 μL of mastermix to each row of a 96-well plate or 12 0.2-mL tubes and place on the gradient block (ensure samples cover a continuous row).
b. PCR program (based on an MJ Research [Reno, NV] gradient block): 93°C 20 min (or appropriate temperature/time to activate *Taq*); 30 cycles of 94°C 30 s, 55–72°C 30 s, 72°C 30 s; then: 72°C 5 min, 4°C storage.

2. The gradient should be visualized on a 1 or 2% agarose gel under trans UV light. The annealing temperature for the assay is based on the brightest single band with the correct amplicon size. Multiple bands and smeared bands that amplify template should not be used because they have the affinity to increase background noise and/or disable the ability to coamplify other DNA regions. There may be recurring band intensity among the samples, in which the highest temperature should be selected to ensure that the individual characteristics of the primers are amplified.

3.4. Performing PCR for Pyrosequencing

1. One of the key issues in this part of the methodology is contamination. Cleaning aides such as 70% ethanol and 5% bleach solution should be used on a regular basis to clean work areas and tools such as pipets before the PCR setup process begins. Also, barrier tips should be used at all times, switching them out when necessary. If a robot mechanism with fixed tips is used, there should be a 5% bleach trough in place to cleanse the tips between every DNA dispensation, as well as between every 96-well plate setup.
2. One microliter of 1–5 ng DNA (amounts may vary from source differentiation) should be aliquoted into the wells of an unskirted 96-well PCR tray (*see* **Note 10**). Every plate should have at least one negative control (no DNA present). In cases where there is more than one primer set per plate, each set should include one negative control to ensure that the primers are not contaminated.
3. Generally, PCR requires a 20-μL reaction for Pyrosequencing. However, the reaction may be reduced to 10 μL in situations where the PCR product is especially strong or wide peaks are indicated in the Pyrogram. An example of the 20 μL based on the ABI Amplitaq Gold PCR mastermix (Applied Biosystems) is as follows: 10 μL ABI Amplitaq Gold PCR mix, forward PCR primer (10 p*M* final concentration), and reverse PCR primer (10 p*M* final concentration) up to 19 μL with 18.2 MΩ-cm water and 1 μL DNA template.
4. Silicon mats or adhesive films should be used to seal the 96-well PCR plate. After the plate is completely prepared the plate should run on a PCR block according to the following protocol (*see* **Note 11**): 93°C 10 min (or relevant temperature/time for *Taq* activation); 55 cycles of 95°C 30 s, X°C 30 s (based on gradient-derived annealing temperature), 72°C 30 s; then 72°C 5 min, 4°C storage.

5. It is useful to check for contamination of the PCR product. Testing a few samples along with the negative control on a 1 or 2% agarose gel will save time and resources, instead of waiting until the Pyrosequencing stage to see if contamination exists. After running the PCR, one may notice condensation on the sides of the wells and on the bottom of the lid. For this reason it is a good idea to centrifuge the plate before removing the lid and continuing the process in order to reduce the possibility of cross-contamination. Five microliters of the product and the negative should be tested on the gel. There only needs to be 10 μL left for Pyrosquencing, so the reduction of volume in the selected test wells should not pose a problem. In some cases, smearing may occur in the PCR product, this may be the result of the many cycles run during the PCR reaction. However, this should not affect the Pyrosequencing reaction as long as there is not any product in the negative.
6. Unless Pyrosequencing takes place directly after the PCR process, the PCR product can be stored at 4°C until needed. Condensation will occur and centrifuging the plate before removing the lid will reduce the chances of contamination. However, if the lid is not properly sealed there is a great chance that the product may evaporate while being stored. The same may also happen if the lid is not properly sealed during the PCR process.

3.5. Preparation of the PCR Product for Pyrosequencing

1. Add the Pyrosequencing primer to the 96-well Pyrosequencing plate (*see* **Note 12**). Cover the Pyrosequencing plate with adhesive film if the setup is to take longer than 15 min to prevent evaporation. These plates can then be stored at 4°C overnight if needed. However, before use, allow to reach ambient temperature and briefly centrifuge.
2. To each well of the 96-well PCR product, add 70 μL sepharose bead mix (**Subheading 2.4.**) and replace lid/adhesive securely.
3. Place the 96-well PCR product plate with bead mix in the plate shaker for 5 min at room temperature (to allow for thorough annealing of the streptavidin-coated sepharose beads to the biotin tag on the PCR primer). Ensure that the plate lid is secure to prevent any cross-contamination between wells (*see* **Note 13**).
4. Set up a workstation for prepping the plates that includes the reagent troughs, PCR product/bead mix tray, and the Pyrosequencing primer tray (**Fig. 2**). Make sure that the PCR product/bead mix plate and the Pyrosequencing primer tray are properly aligned so that the negatives are in the same orientation before transferring the samples. Aliquot reagents into troughs.
5. Shake the vacuum tool (turned off) in clean 18.2 MΩ-cm water to release any beads or debris that may be on it (*see* **Note 14**). Discard the remaining water, refill the trough, and turn the vacuum on. Leave the vacuum tool in the trough until all water has been removed (approx 30 s).
6. Place the filter tips of the vacuum tool into the wells of the PCR/bead mix plate and let it remain until all liquid has been removed from the plate (*see* **Note 15**).

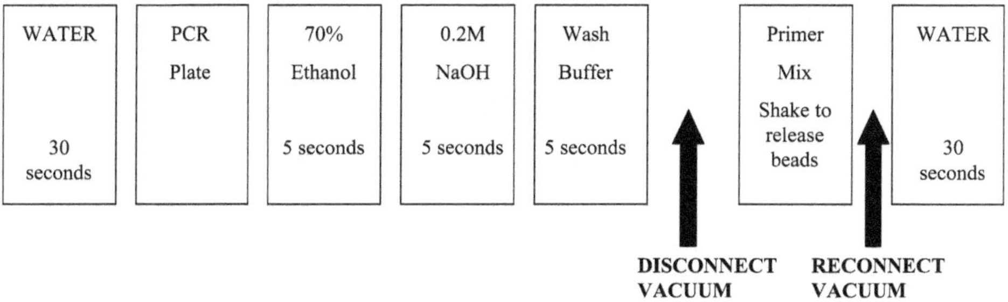

Fig. 2. The Pyrosequencing® reagent set up for processing the PCR and Pyrosequencing plates.

7. Place the vacuum in the 70% ethanol trough. When the liquid begins to flow through the tubing, allow the filter tips to suck up the ethanol for 5 s.
8. Repeat **step 7** for the 0.2 *M* NaOH (denatures the DNA to single-stranded PCR) and washing buffer (cleans and neutralizes the PCR product).
9. Disconnect the vacuum hose from the vacuum tool and place the vacuum tool into the Pyrosequencing plate containing the Pyrosequencing primer/annealing buffer mix. If the vacuum hose is still connected when placed into the Pyrosequencing plate, the primer mix will be sucked through the tips causing you to lose your PCR product. Gently shake/rock the tips of the vacuum in the wells of the Pyrosequencing plate to disperse the PCR product (*see* **Note 16**).
10. Remove the vacuum tool from the Pyrosequencing plate once the shaking/rocking is complete. Reconnect the vacuum tool to the hose and place the tool into clean 18.2 MΩ-cm water to cleanse it for the next plate.
11. Place the Pyrosequencing plate on the heat block for 2 min at 80°C (*see* **Note 17**). After 2 min, remove the plate from the heat block and place on a surface to cool. Once cooled, an adhesive film can be used to cover the plate (unless the plate will be run within 15 min) to prevent evaporation (*see* **Note 18**).
12. Processed plates may be stored at 4°C until needed.

3.6. Pyrosequencing

3.6.1. Entering Assay Details

1. To open the Pyrosequencing software, a user name and password is required that is typically set up during instrument installation. Individual or group-wide passwords may be used.
2. To enter an assay that is not already entered into the software, click "Simplex Entry" on the left side of the screen (*see* **Note 19**). Right-click "Simplex Entry" in the menu tree (located to the right of the Simplex Entry icon) and select "New Entry" (**Fig. 3**).
3. Enter a unique name for the assay (typically the gene/SNP name), as well as a sequence to analyze, which includes 5–6 bases after the SNP position. SNPs should be denoted as:

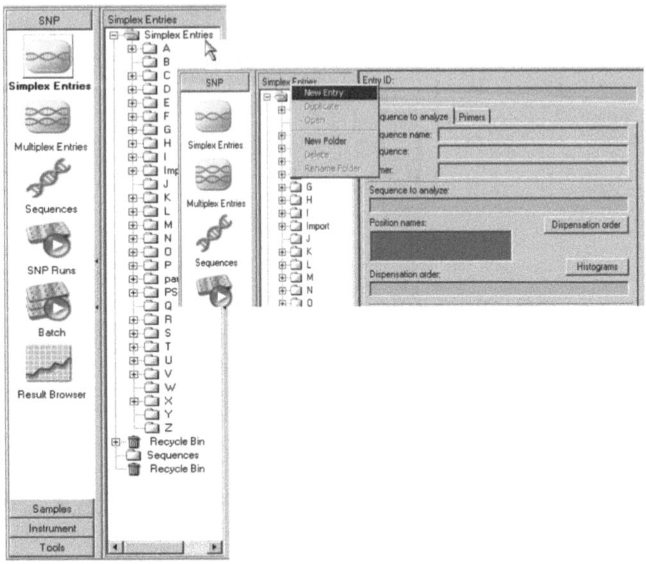

Fig. 3. Example of how to enter a Simplex Entry.

SNP	T/C
Tri-allelic SNP	T/C/A
Tetra-allelic SNP	T/C/A/G
Indels	[T]
Short repeats	[TA][TA][TA]

To generate the least amount of nucleotide dispensations for optimum genotype information, click "Dispensation Order." For troubleshooting purposes, the dispensation order can also be manually edited by typing in the field.

4. To generate the predicted dispensation pattern, select "Histograms." Both homozygous and heterozygous patterns will be displayed. If using multiplex, one may scroll through the histograms on the lower panel for viewing purposes. Click "Export" (near the bottom of the screen) to open the histograms in a browser window to print or save any/all histograms provided.
5. Click "Save." After saving, if any alterations need to be made, a duplicate setup with another unique name must be made, for the parameters can no longer be changed.

3.6.2. Entering a SNP Run

1. Click the "SNP Run" icon to the left of the screen.
2. Right-click "SNP Run" on the menu tree, and select "New SNP Run." The menu tree for "SNP Runs," as well as "Batch Runs," may be organized into multiple folders so that various users may have access to their own files. Once a relevant folder has been created, one can simply right-click over it and select "New Run."

3. Enter a unique "Name" (e.g., Gene/SNP/Sample Set/Date). Select the appropriate "Instrument Parameters" from the drop-down menu. The instrument parameters are typically a default file, however make sure that appropriate parameters are selected for the capillary- or nucleotide-dispensing tips (not interchangeable). The parameter setup instructions are found in the dispensing tip packages. The default plate map is for a full 96-well plate. If only a partial plate is used, individual wells may be selected by holding down the control key and clicking on the selected wells. Click "Activate Wells" and the unused wells will turn gray while all active wells remain white.
4. Under the Setup tab, select a SNP assay by clicking on the drop-down menu for Simplex Entry and selecting the name entered in **Subheading 3.6.1.** Once selected, click and drag over all active wells for the designated assay. If there is more than one assay for a plate, simply repeat this step until all active wells have a designated assay.
5. Click "Save." The plate may be edited after being saved, and any changes may be saved as well.
6. For multiple plates using the same assay, right-click over the assay just entered (in the "SNP Runs" file), and select "Duplicate SNP Run." Only a unique run name is all that is required.

3.6.3. Individual Plate Run for PSQhs96 and PSQhs96A

1. From the SNP run setup in **Subheading 3.6.2.**, click the "View" tab and then select "Run." This page lists the appropriate volumes of nucleotides, enzyme, and substrate needed for the run.
2. Clean both the nucleotide/capillary and reagent tips before use (*see* **Note 20**). Fill the tips with 18.2 MΩ-cm water and apply pressure over the top of the tip to check for any tip blockages. If water does not squirt from the bottom of the tip, empty/refill several times to try to force the water through. If the tip remains blocked, discard and get new tips (make sure to label the tips accordingly). New tips do not need cleaning. Set up the cartridge as shown in **Fig. 4**.
3. Nucleotides, enzyme, and substrate (clearly labeled vials and bottles) are sold in reagent kits. Nucleotides come as a solution, however enzyme and substrate are lyophilized and should be resuspended with 18.2 MΩ-cm water before use (amount is on the label). The enzyme and substrate rapidly dissolve (no shaking or mixing required). If shaken, air bubbles could result causing tip blockages or inconsistent dispensation. Unused resuspended enzyme and substrate can be stored at –20°C for future use. For all dispensing tips it is recommended that nonbarrier pipet tips are used, as fiber can cause tip blockages.
4. For nucleotide-dispensing tips, microfuge the nucleotide vials for 10 min and *do not* aliquot from the bottom of the vial (precipitate may also cause tip blockage). In addition, the tips should be filled by *doubling* the amount of nucleotide suggested by the software. For capillary-dispensing tips, in a microfuge tube, make a 1:1 dilution with the nucleotides and TE buffer pH 8.0. Mix well before use.

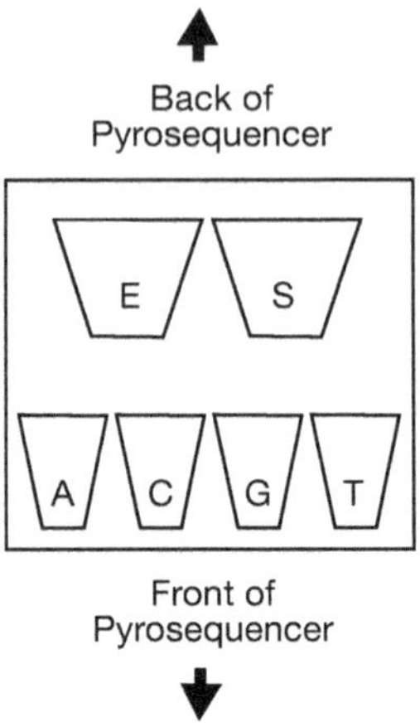

Fig. 4. The Pyrosequencing® cartridge containing the reagents and nucleotides. E, enzyme; S, substrate; nucleotides, A, C, G, T.

5. Fill the nucleotide and reagent tips with the appropriate volumes according to the amounts suggested by the software. Gently dispense liquid down the sides of the tips to prevent pipetting air bubbles into them and causing blockages. Be sure to check for any air bubbles in the nucleotide-dispensing tips. If any air bubbles are present, simply tap the sides of the tips until the air bubbles surface, or, dislodge them with a clean pipet tip.
6. Run a test plate after filling the cartridge. Place the cartridge in the Pyrosequencing instrument and the test plate in the 96-well plate platform (*see* **Note 21**). Select the "Instrument" tab on the left side of the screen, then click "Manage." Select the instrument from the drop-down menu. Click "Test," and a warning will appear asking you to check that the test plate has been placed in the instrument. Click "Ok." When using nucleotide-dispensing tips, at least three to four test plates should be run to ensure no blockage has occurred. The test takes approx 30 s. Once complete, remove the test plate, and in the middle of the plate there should be liquid dots over six of the wells representing: four nucleotides, an enzyme, and a substrate. If there are less than six small dots, a blockage has occurred and the tips should be checked and removed of any blockages.
7. Remove the adhesive film from the Pyrosequencing plate (if applied) and place the plate in the Pyrosequencing 96-well plate platform. Close all levers, select the instrument from the drop-down menu, and click "Run" on the individual plate run setup. The enzyme, substrate, and nucleotide will dispense in the predetermined order. Click on any of the wells to view the progress of the run.
8. Click on "Analyze All" for the program to automatically analyze the data once the run is complete.

3.6.4. Batch Runs Using the PSQhs96A

1. The setup for SNP runs should be performed as in **Subheading 3.6.2.** A maximum of 10 plates may be run per batch.
2. Once runs are saved and closed, on the left side click "Batch." On the menu tree, right-click "Batch Runs" and then "New Batch." Use a new and unique name for the batch, select the proper instrument parameters, and uncheck the "Barcode" box if barcodes are not used.
3. Click "SNP Runs" on the left side of the screen. From the menu tree, click and drag all of the runs for the batch into the batch window. The order may be edited by right-clicking on a particular run, and options such as "Move Up, Down, and Remove" appear.
4. At the top of the screen, select "Batch" and "Setup Information." A browser window will open providing the total amount of enzyme, substrate, and nucleotides (the amount should be *doubled* for the capillary-dispensing tips) for the entire batch.
5. Clean the dispensing tips and set up the cartridge as described in **Subheading 3.6.3.**, and run a test plate.
6. Remove all adhesive film from the Pyrosequencing plates, stack them in the proper order (as listed in the batch run), and place in the stacker unit (*see* **Note 22**). The proper plate orientation is shown at the top of the stacker unit, and plate one of the batch should go at the top with the last plate at the bottom of the stacker.
7. Firmly push the stacker unit into the Pyrosequencing instrument (the nucleotides will not dispense if it is not completely in the machine).
8. Click the "Play" icon and the plates will automatically load and be discarded throughout the run. In addition, the plates will automatically be analyzed by the software at the end of each run. When each run is complete, they can be accessed from both the "Batch Run" and the "SNP Run" files.

6.3.5. Analysis of Pyrosequencing Results

1. **Figure 5** shows examples of Pyrogram charts and predicted histograms for a SNP, a tri-allelic, and an indel polymorphism. Orange wells that call for human intervention may be edited by clicking on the well of interest, and opening the predicted histograms from the "Histograms" tab on the right side of the screen. A genotype may be passed/failed/checked, as well as the genotype itself changed appropriately, according to the predicted histogram by right-clicking over the genotype above the Pyrogram. Once a well has been edited (in any way), a dark circle will show on the plate map to show it has been edited.
2. Negative controls (any wells without DNA present) should automatically be scored as negative. There may be nonspecific peaks in the negative control wells, however it is typically caused by looping of the internal primer. This may aid in troubleshooting assays because it may help to determine if the internal primer is the cause of background peaks.

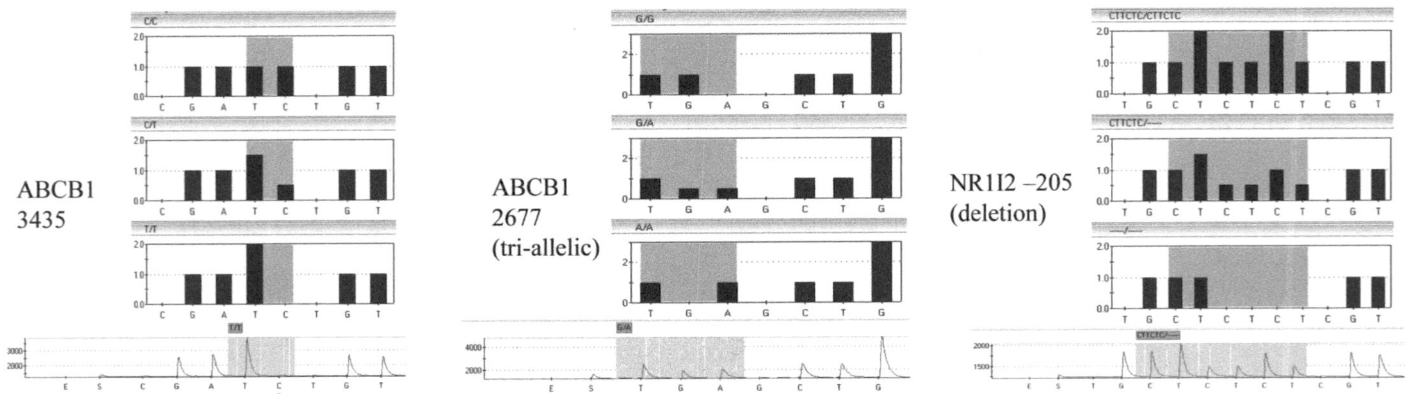

Fig. 5. Some of the predicted histograms and Pyrogram® charts for a single-nucleotide polymorphism (*ABCB1* 3435), tri-allelic (*ABCB1* 2677 G>A/T), and indel (*NR1I2* -205).

3. The data can be exported as a report, a tab-delimited file, or an XML file (for whole plates and/or selected wells). Custom export options are also available. To export a file, click "Report" and then save as the appropriate file type. The report structure is available in forms that transfer to most database and spreadsheet systems. Pyrogram charts may also be saved and printed.

4. Notes

1. DNA extraction kits (e.g., Gentra and Qiagen) do not inhibit the assay.
2. A test prior to performing PCR should be done with dilutions of DNA template and PCR primers to establish the optimum concentration of DNA that can produce a clean, high-yield PCR product. The quality of the DNA depends on the extraction and the source from which it is derived. For example, DNA extracted from blood may require a minimum of 0.5–1.0 ng to produce a reliable product. However, if using DNA from plasma, serum, frozen tissue, or whole genome-amplified products, an increased amount of template may be required as a result of the susceptibility of fragmentation.
3. To minimize pipetting errors and to provide consistent results, premade PCR mastermixes (containing buffer, magnesium, dNTPs, and *Taq* polymerase) are highly recommended.
4. Hot start is recommended, however non-hot start *Taq* may be used as well (although chances of primer dimer formation may increase).
5. The vacuum tool should be stored dry and may be stored in an empty tip box to help prevent any dust/debris from getting on the filter tips.
6. Some software may offer a secondary structure to check for template looping, which is when the primers or the single-stranded amplicon, folds on top of itself and does not allow the polymorphism to be amplified. Even if the software does not allow for a secondary structure to be seen, one should try to circumvent template looping at all cost.
7. For multiplexing, up to three internal primers may be created from either the same PCR product or different PCR products. With only one internal primer designed at a time, often the first-choice primers for each will not be useful in a multiplex assay (where the combined sequence to analyze is best designated to generate unique SNP dispensations).
8. For Pyrosequencing primers that have flagged critical scoring parameters, it is possible to "trick" the software to improve the search. Enter a "fake" SNP either five bases before or after the SNP (the software only examines five bases on either side of the SNP for a primer) to extend the region searched, which helps to overcome mispriming and dimer problems.
9. Adjusting the PCR primers at the 5'-end may help minimize template loops, e.g., shifting the primer two to three bases (or trying in a different location in general).
10. For assays that may require a larger volume of DNA, a reduction in the volume of water for the PCR mastermix may help optimize the PCR assay. In addition, DNA may be dispensed and dried overnight at room temperature into a plate for later use. Extra water should be added to the PCR mastermix to compensate for the decrease in volume. For a 20-μL PCR reaction, 100 μL more of 18.2 MΩ

water should be added (instead of 960 μL, it would be 1060 μL of water), and thus 20 μL of the mastermix would be dispensed into each well. The DNA is reconstituted once the PCR mastermix is added.

11. For PCR, 55 cycles are run to ensure that all of the primers and nucleotides are exhausted to prevent background noise for Pyrosequencing.
12. For one 96-well Pyrosequencing plate: add 43.2 μL of the internal primer to 1396.8 μL annealing buffer. Dispense 12 μL in each well. In addition, multiple assays can be run on a single 96-well Pyrosequencing plate. Internal primer should be used for the negative controls from the corresponding PCR plates for each assay used for trouble-shooting program background issues, as well as controls.
13. If using the Eppendorf thermomixer, 1400 rpm is the optimum speed. Also, this plate is to be used immediately. If allowed to sit, the beads will settle at the bottom of the wells and will not be accessible to the vacuum tool. If this has occurred, place the plate back into the shaker for 1–2 min to redisperse the beads.
14. Always check the filter tips prior to using them. If there seems to be a build up of beads on the tips that does not release upon cleaning, place the vacuum in a sonicator to help release the beads and further clean the tips.
15. Gentle rocking of the vacuum tool may be used to prevent surface tension, causing the liquid to remain in the wells. The PCR product will not pass through the filters because of the beads that are now attached to the biotin primer. In addition, if the filter tips have turned gray, the vacuum is taking a long time to suck from a trough or certain wells from a Pyrosequencing plate are not being sucked up, and the filter tips may need to be changed (check the vacuum pressure to ensure it is at maximum).
16. If vacuum is not rocked properly, a pattern of failed wells around the edges of your plate after it has been sequenced may result.
17. Ensure the Pyrosequencing plate rests completely on the Pyrosequencing plate adaptor with the lid is placed completely over the plate to prevent any evaporation.
18. If evaporation had occurred, add 12 μL of annealing buffer if evaporation has occurred, so the plate may be reused. On the other hand, if the Pyrosequencing plate is covered while the plate is too warm, condensation will form on the lid, which can lead to cross-contamination of the wells upon removal.
19. For multiplex assays, click the "Multiplex Entry" icon, right-click "Multiplex Entry" on the menu tree, select "New Entry." Type the three dispensation orders for each Pyrosequencing primer. Combined dispensation order is computer generated.
20. For storage, keep clean tips in a clean, lint-free environment (e.g., the storage box that is provided with the instrument).
21. Place adhesive film over the test Pyrosequencing plate so that the dispensation of the reagents and nucleotides may be easily seen, which can then be wiped off once complete and reused.
22. Be careful to remove all adhesive film from the Pyrosequencing plates and make sure that the plates can be easily lifted without sticking to any other plates. If there is any residual film, the robotic arm may jam because of the plates sticking together.

Acknowledgments

The authors are supported by U01 GM63340.

References

1. Sachidanandam, R., Weissman, D., Schmidt, S. C., et al. (2001) A map of human genome sequence variation containing 1.42 million single nucleotide polymorphisms. *Nature* **409,** 928–933.
2. Ahluwalia, R., Freimuth, R., McLeod, H. L., and Marsh, S. (2003) Use of pyrosequencing to detect clinically relevant polymorphisms in dihydropyrimidine dehydrogenase. *Clin. Chem.* **49,** 1661–1664.
3. Garsa, A. A., McLeod, H. L., and Marsh, S. (2005) CYP3A4 and CYP3A5 genotyping by Pyrosequencing. *BMC Med Genet.* **6,** 19.
4. Mathijssen, R. H., Marsh, S., Karlsson, M. O., et al. (2003) Irinotecan pathway genotype analysis to predict pharmacokinetics. *Clin. Cancer Res.* **9,** 3246–3253.
5. Marsh, S., King, C. R., Garsa, A. A., and McLeod, H. L. (2005) Pyrosequencing of clinically relevant polymorphisms. *Methods Mol. Biol.* **311,** 97–114.
6. King, C. R., Yu, J., Freimuth, R. R., McLeod, H. L., and Marsh, S. (2005) Interethnic variability of ERCC2 polymorphisms. *Pharmacogenomics J.* **5,** 54–59.
7. Rose, C. M., Marsh, S., Ameyaw, M. M., and McLeod, H. L. (2003) Pharmacogenetic analysis of clinically relevant genetic polymorphisms. *Methods Mol Med.* **85,** 225–237.

5

Universal Primer Applications for Pyrosequencing®

Dong-Chuan Guo and Dianna M. Milewicz

Summary

Pyrosequencing® is a high-throughput technique for single-nucleotide polymorphism (SNP) genotyping, DNA sequencing, and SNP allele frequency and DNA methylation assays. The Pyrosequencing assay involves initially purifying PCR-generated single-strand DNA labeled with biotin as a sequencing template. The disadvantage of the synthesis of a biotinylated sequence-specific primer to assay each DNA variant lies in costs and time consumption. To overcome this problem, methodology has been developed to generate biotinylated Pyrosequencing DNA templates using a universal biotinylated primer. Using this methodology, biotinylated DNA fragments can be generated without the use of sequence-specific biotinylated primers for each DNA variant.

Key Words: Biotinylated DNA fragment; Pyrosequencing; single-nucleotide polymorphism; SNP; universal primer.

1. Introduction

Pyrosequencing® is a high-throughput real-time DNA sequencing technique for single-nucleotide polymorphism (SNP) genotyping, DNA sequencing, SNP allele frequency, and DNA methylation assays. This technique has been extensively applied for population genetics, medical genetics, pharmacogenomics, and forensic DNA analysis *(1,2)*. An advantage of the Pyrosequencing SNP assay is that it not only detects the SNP but also a few base pairs of flanking DNA sequence that can serve as internal controls for monitoring the specificity of SNP assays.

The Pyrosequencing technique uses a nonelectrophoresis-based bioluminometric DNA sequencing method that employs a cascade of luciferase-based enzymatic systems to monitor DNA synthesis in real time. The assay is performed at 28°C because of the low thermostability of the firefly luciferase involved in generating a visible light signal during DNA synthesis.

From: *Methods in Molecular Biology, vol. 373: Pyrosequencing® Protocols*
Edited by: S. Marsh © Humana Press Inc., Totowa, NJ

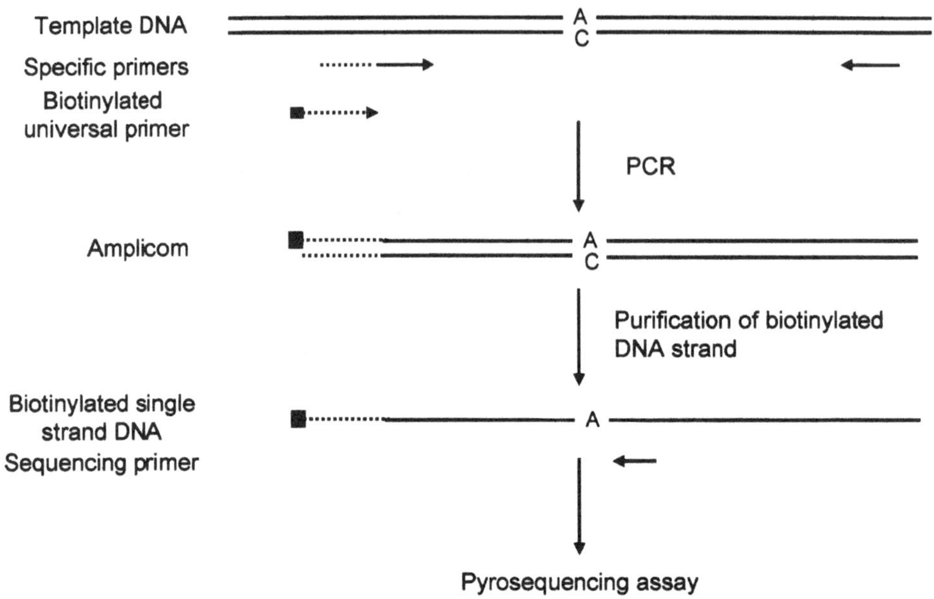

Fig. 1. Schematic representation of the nested PCR methods used to amplify a genomic variation using a biotinylated universal primer.

Performing real-time DNA sequencing at room temperature may result in background noise because of nonspecific binding of the sequencing primer to the template or the formation of 3'-end loops. To overcome this problem, single-strand DNA is used as a template to increase the annealing specificity of the sequencing primer to the template *(3)*.

The Pyrosequencing assay involves initially purifying PCR-generated single-stranded DNA as a sequencing template and then annealing with a sequencing primer to perform real-time DNA sequencing. In order to purify a single-stranded DNA template, a biotin-labeled DNA fragment is generated by PCR amplification using a sequence-specific biotinylated PCR primer. Biotin has a high affinity for streptavidin, therefore, the biotinylated template can be immobilized to streptavidin-coated beads and purified after denaturation of the double-stranded PCR product with 0.2 *N* NaOH. The disadvantage of this methodology lies in the cost and time associated with the synthesis of an end-labeled, sequence-specific primer to assay each DNA variant.

A methodology of universal biotinylated primers has recently been developed to significantly reduce the cost of generating biotinylated Pyrosequencing templates *(4–6)*. This methodology uses a universal biotinylated primer instead of sequence-specific biotinylated PCR primers to generate biotinylated amplicons. **Figure 1** depicts the strategy for generating labeled amplified DNA fragments using the universal primer. In this process, three primers are utilized including a biotinylated universal primer and two sequence-specific primers. One of the sequence-specific primers contains a 5'-tail that is complementary to the universal primer. During PCR amplification, the sequence-specific primer pair initiates PCR amplification of the genomic DNA. These initial amplified fragments then serve as templates for the universal biotinylated primer and a sequence-specific primer in subsequent amplifications, thus producing labeled amplicons.

By using universal primers, alternative chemically tagged labeled DNA fragments can be generated without the requirement of labeling sequence-specific primers. Only one universal biotinylated primer is enough to PCR amplify all different genomic variants analyzed by Pyrosequencing, which significantly reduces the expense and time traditionally involved in synthesizing sequence-specific labeled primers. In addition, this methodology can also be applied to generate other chemically tagged labeled DNA fragments, such as fluorescently labeled DNA fragments *(4)*.

2. Materials

1. Human genomic DNA was extracted from peripheral blood samples using the PureGene genomic DNA isolation kit (Gentra Systems, Minneapolis, MN). Genomic DNA samples tested include two homozygotes, A or G, and one heterozygote, A/G, of SNP rs243834 in the *matrix metalloproteinase-2* (*MMP2*) gene.
2. Primers: universal biotinylated primer: 5' biotin-ATCTGTGCCGAGGCTCAGGC.
   Tailed sequence-specific primer:
   5' GTGCCGAGGCTCAGGCTGCTGGTTCACTGTGTCTG.
   Sequence-specific primer: 5' ACTTGGGAAAGCCAGGATCCA.
   Pyrosequencing primer: 5' CTTCACCTCATTGTATC.
3. HotStart Taq™ DNA polymerase (Qiagen Inc., Valencia, CA).
4. Streptavidin-sepharose high-performance beads (Amersham Biosciences, Buckinghamshire, UK).
5. Pyro Gold Reagent Kit (Biotage, AB, Uppsala, Sweden) and other reagents for the Pyrosequencing assay.

3. Methods

The methodology for using universal primer to generate biotinylated amplicons involves a single-step nested PCR approach using a universal and two sequence-specific primers (*see* **Note 1**). In order to reduce the background and increase the efficiency of the PCR amplification, the nucleotide sequence of the universal biotinylated primer is designed to have the following elements: no homology to the human genome sequence in its 3'-terminus (*see* **Note 2**), a length of 20 bp, and a G-C content of around 60%. The tailed sequence-specific primer contains a 5'-tail with 16 nucleotides complementary to the 3'-terminus of universal biotinylated primer.

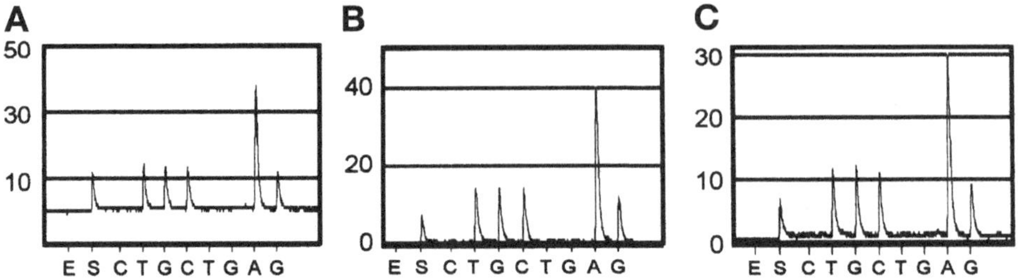

Fig. 2. Results of the Pyrosequencing® assay A/G polymorphism in *MMP2* used DNA templates that were generated by a biotinylated sequence-specific primer and a biotinylated universal primer. **(A)** DNA template generated using a biotinylated sequence-specific primer. **(B,C)** DNA template generated using a biotinylated universal primer with concentration of tailed sequence-specific primer at 25 and 5 f*M*, respectively. Because single-nucleotide polymorphism (SNP) was tested on the reverse DNA strand, the SNP is shown in the figure as T/C.

To optimize the conditions for obtaining high yields of biotinylated amplicons, the concentration and ratio of universal and sequence-specific primers have been tested. The concentration of the tailed specific primer is critical because it dictates the efficiency of producing a higher ratio of biotinylated vs nonbiotinylated amplicons. In the optimization tests, the concentration of the labeled universal primer and the sequence-specific primer were 0.5 p*M*, whereas the concentration of the tailed specific primer varied between 1 and 100 f*M*. Final results show that all of the biotinylated DNA fragments generated using different concentrations of the tailed specific primer yield high-quality Pyrosequencing signals (**Fig. 2**). This indicates that use of the universal biotinylated primer methodology can produce biotinylated amplicons within a wide-concentration range of the tailed specific primer. We selectively used 25 f*M* of tailed specific primer for generating biotinylated amplicons for the Pyrosequencing assay *(4)*.

The methodology described has been used for SNP genotyping using a PSQ™ 96 Pyrosequencing instrument (**Fig. 3**); however, it can also be applied for DNA sequencing, allele frequency assays, and DNA methylation analysis with other Pyrosequencing platforms.

3.1. Designing the Pyrosequencing Primers

1. PCR primers were designed for each of the SNPs using the Primer3 program (http://www.genome.wi.mit.edu/cgi-bin/primer/primer3_www.cgi).
2. The DNA sequence of amplicons designed by the Primer3 program is input into the SNP Primer Design Software (http://techsupport.Pyrosequencing.com/) (Biotage AB) for designing the Pyrosequencing primer, which can be complementary to either the forward or reverse DNA strand (*see* **Note 3**). After the primer sequence is chosen, the sequence of the universal primer is added on the 3'

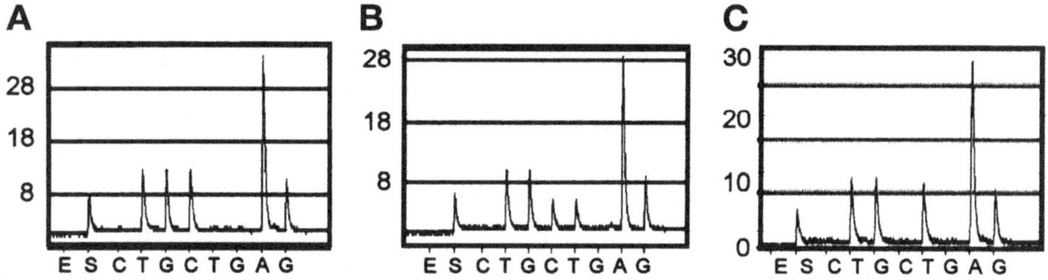

Fig. 3. Results of the Pyrosequencing® assay A/G polymorphism in *MMP2* using DNA templates that were generated by a biotinylated universal primer. Three different genotypes can be identified using DNA templates generated by the methodology of universal primer. **(A)** GG genotype, **(B)** A/G genotype, and **(C)** AA genotype.

terminus of the template strand and reanalyzed with the SNP Primer Design Software to ensure that there are no 3'-end loop formations from the addition of the universal primer sequence. All of the primers are synthesized by Integrated DNA Technologies (Coralville, IA).

3.2. PCR Amplification

1. PCR was carried out by using HotStart Taq™ DNA polymerase (Qiagen Inc.). The recipe of each PCR mixture is:

DNA (5–15 ng/μL)	1.0 μL
5X Q buffer	2.0 μL
10X PCR buffer	1.0 μL
dNTP (2.5 m*M*)	0.8 μL
$MgCl_2$ (25 m*M*)	0.5 μL
Universal biotinylated primer (10 p*M*/μL)	0.5 μL
Tailed sequence-specific primer (0.5 p*M*/μL)	0.5 μL
Sequence-specific primer (10 p*M*/μL)	0.5 μL
H_2O	3.2 μL
HotStart Taq DNA polymerase (5 U/μL)	0.06 μL

2. PCR amplification was carried out with a two-cycle amplification program. Conditions for PCR consist of an initial step of 15 min at 95°C, for activating HotStar *Taq* DNA polymerase. The first cycle program includes 10 cycles of 95°C for 30 s; 60°C for 30 s; and 72°C for 30 s. The second cycle program includes 35 cycles of 95°C for 30 s; 54°C for 30 s; and 72°C for 30 s, followed by a final 5-min extension step at 72°C.

3.3. Pyrosequencing Assay

1. Immobilize biotinylated PCR product to beads: add 30 μL H_2O, 3 μL streptavidin-sepharose beads, and 40 μL binding buffer into 10 μL biotinylated PCR product and mix in a shaker at room temperature for 10 min.

2. Purify biotinylated single-strand DNA: capture the beads on the filter probes by using a Vacuum Prep Tool. Move the Vacuum Prep Tool to the troughs with 70% ethanol, 0.2 *N* NaOH denaturation solution, and washing buffer, and let each solution flush through the filters for 5 s.
3. Anneal the Pyrosequencing primer with the single-strand DNA template: release the beads in a PSQ 96-Plate Low, prefilled with 0.3 μ*M* sequencing primer in 40 μL annealing buffer. Heat the plate at 80°C for 2 min and then cool to room temperature to allow the Pyrosequencing primer to anneal with the single-strand DNA template.
4. Perform SNP analysis: SNP assays are performed using the conditions recommended by the manufacturer on a Pyrosequencing instrument with PSQ™ 96 SNP Reagent Kit (Biotage AB).

4. Notes

1. Because a biotinylated universal primer can be applied for PCR amplifications of different DNA fragments, a large quantity of biotinylated universal primer can be ordered, which will further decrease the expense and time involved in synthesizing the universal primer.
2. Because a universal primer is used for different PCR amplifications, it is important to avoid cross-contamination during PCR amplification.
3. Designing a short amplicon as template for a Pyrosequencing assay can reduce background noise of the Pyrosequencing assay and increase PCR efficiency.

Acknowledgments

This work was support by MO1RR02558 (General Clinical Research Center), RO1 HL62594, R01 AR46718, and Doris Duke Foundation.

References

1. Langaee, T. and Ronaghi, M. (2005) Genetic variation analyses by Pyrosequencing. *Mutat. Res.* **573,** 96–102.
2. Diggle, M. A. and Clarke, S. C. (2004) Pyrosequencing: sequence typing at the speed of light. *Mol. Biotechnol.* **28,** 129–137.
3. Ronaghi, M. (2001) Pyrosequencing sheds light on DNA sequencing. *Genome Res.* **11,** 3–11.
4. Guo, D. C. and Milewicz, D. M. (2003) Methodology for using a universal primer to label amplified DNA segments for molecular analysis. *Biotechnol. Lett.* **25,** 2079–2083.
5. Schuelke, M. (2000) An economic method for the fluorescent labeling of PCR fragments. *Nat. Biotechnol.* **18,** 233–234.
6. Pacey-Miller, T. and Henry, R. (2003) Single-nucleotide polymorphism detection in plants using a single-stranded Pyrosequencing protocol with a universal biotinylated primer. *Anal. Biochem.* **317,** 166–170.

6

Allele Quantification and DNA Pooling Methods

Jonathon Wasson

Summary

Studies utilizing differences in single-nucleotide polymorphism allele frequencies between cases and controls have been widely used in genetic analyses to locate putative genes or chromosomal regions that may be associated with a disease. In these studies the assessment of allele frequencies can be expedited and the genotyping costs reduced by the use of DNA pools. There have been multiple studies that have reported the accuracy of Pyrosequencing® for the assessment of allele frequencies in DNA pools. In addition, there are an increasing number of other types of studies that make use of allele quantification to evaluate a disease status or to make a clinical diagnosis. In this chapter, the making of DNA pools is described, as well as the use of Pyrosequencing to quantify alleles. The ease of use, short run, and analysis times make Pyrosequencing the preferred method.

Key Words: Pyrosequencing®; Pyrogram®; DNA pools; single-nucleotide polymorphism; SNP; allele frequency.

1. Introduction

Mapping the genetic factors involved in common diseases, such as heart disease, cancers, diabetes, and a number of autoimmune conditions, is an area of enormous interest in both academic and commercial sectors. Linkage analysis in families with multiple affected individuals has been the standard method for discovering Mendelian disease genes *(1)*, however, for most complex genetic diseases a single gene is neither necessary nor sufficient for the disease etiology. More recently, the discovery process for finding complex disease genes has been by the use of single-nucleotide polymorphism (SNP)-association mapping *(2)*. This is a powerful approach for fine-scale localization of genetic variants conferring increased susceptibility to complex disease, and it has been shown that there is increased power for allelic association over linkage to detect disease determinants of relatively small effect *(3)*. Identification of putative

From: *Methods in Molecular Biology, vol. 373: Pyrosequencing® Protocols*
Edited by: S. Marsh © Humana Press Inc., Totowa, NJ

susceptibility genes for type 2 diabetes *(4)*, Crohn's disease *(5)*, asthma *(6)*, and schizophrenia *(7)* are successful examples of SNP-association mapping. Thus, many seek rapid and cost-efficient methods for this approach. Since SNP genotyping technology first emerged, individual genotyping costs have dropped dramatically, from $1 a piece to just 10 cents or less in some methods depending on the throughput. This can still be a costly enterprise when dealing with the thousands of samples and the thousands of SNPs that these association studies often require. Pooling the DNA of theoretically up to thousands of individuals into just one sample reduces the costs and labor dramatically. A recent SNP workshop *(8)* reported that the use of DNA pools for screening many samples for differences in allele frequencies at many loci was not sufficiently reliable for use on a genome-wide scale at this time, however, two recent successful studies show that using the pooled DNA strategy works *(9,10)*.

There is substantial interest in allele quantification in other scientific fields. It is being used for the detection of chromosomal aneuplodies, which are a common cause of congenital disorders associated with cognitive impairment and multiple dysmorphic features *(11)*, for the assessment of methylation of CpG sites in cancer that can direct a patient's treatment *(12)*, and for the quantification of allele-specific G protein mRNA transcripts that serve as a pharmacogenetic marker for multiple drugs *(13)*. The materials and methods described here can also be used in these applications.

2. Materials

2.1. DNA Isolation

1. Puregene kit (Gentra Systems, Minneapolis, MN).

2.2. DNA Quantification

1. Quant-iT DNA Assay Kit (Molecular Probes, Eugene, OR).
2. Fluorescent plate reader.

2.3. Construction of DNA Pools

1. Sterile 55-mL polypropylene solution basins (Labcor Products, Frederick, MD).
2. Accurately calibrated multichannel pipettors.

2.4. Amplification of SNP by PCR

1. Amplitaq Gold PCR Master Mix (Applied Biosystems, Foster City, CA).
2. PCR primers: one primer (depending on the SNP sequencing primer; *see* **Subheading 3.4., step 2**) with 5' tag: AGC GCT GCT CCG GTT CAT. Universal biotin-labeled primer: 5'TEG Biotin—AGC GCT GCT CCG GTT CAT AGA TT. SNP sequencing primer as designed by the web-based software at http://techsupport.Pyrosequencing.com (*see* **Note 1**) (IDT, Coralville, IA).
3. Chimney top PCR plates (ISC Bioexpress, Kaysville, UT).

2.5. Template Isolation

1. Dynabeads M-280 strepavidin (Dynal Biotech, Oslo, Norway).
2. Hydra Robotic microdispensors (Robbins Scientific, Sunnyvale, CA) or other robots.

2.6. Pyrosequencing

1. Reagents and plastic ware (Pyrosequencing, Biotage AB, Uppsala, Sweden).

2.7. Allele Quantification

1. AQ software (Pyrosequencing, Biotage AB).

3. Methods

The critical parameters in allele quantification include the sensitivity of the quantification software, the accuracy of the DNA pool construction, and the adjustment of the data to compensate for possible irregularities in the Pyrogram®. In terms of the sensitivity of the software, several papers have reported a significant degree of correlation between allele frequencies and Pyrogram peak heights, with reported R^2 statistics of 0.981, 0.993, and 0.9963 ***(14–16)***. These groups also reported significant concordance between the allele frequencies for multiple SNPs as measured in pools of various sizes vs as measured from individual genotypes: the deltas ranged from 1.1 ± 0.6% ***(16)*** to 2.37 ± 0.11 ***(14)***. The size of this delta is influenced by the accuracy of the measurement of the DNA concentration of the samples contained in the pools and by the calibration of the pipets used to make the pools. Preferential allele amplification, an irregularity in the PCR, and the increased nucleotide signal in the Pyrosequencing, can lead to skewing of the SNP nucleotide peak heights in the Pyrogram and will cause inaccurate allele quantification if not detected and adjusted for in the analysis.

3.1. Isolation of Genomic DNA From Whole Blood

1. Perform as described by Puregene kit. Working DNA stored at 4°C and stocks at –20°C (*see* **Note 2**).

3.2. DNA Quantification on Fluorescent Plate Reader

1. Perform as described by Quant-iT kit (*see* **Note 3**).
2. Dilute working samples with high-purity water to a concentration of 10 ng/μL in sterile 1.2-mL polypropylene tubes, arranged in 96-well tube format racks, and stored at 4°C in a dedicated refrigerator separate from PCR products.

3.3. Construction of DNA Pools

1. Allow the samples to come to room temperature.
2. Gently mix the 96-well format racked DNA samples on a rocking platform to ensure homogeneity before pipetting.
3. Spin the rack to prevent spillage before opening.

4. Pipet, using an accurately calibrated multichannel pipet, equal volumes of each sample into a sterile 55-mL polypropylene solution basin. Once the individual samples for that particular pool are dispensed, rock the basin containing the DNA (carefully to avoid spillage) for several minutes to do a preliminary mixing (*see* **Note 4**).
5. Pipet the DNA from the basin into a 50-mL polypropylene tube, making every effort to recover all liquid in the basin.
6. Rock the tube for approx 1 h.
7. Pipet the pooled DNA into 1-mL aliquots in sterile 1.5-mL polypropylene microtubes and store at 4°C in the dedicated refrigerator.
8. Wrap the tops of stored tubes with Parafilm.
9. Conduct quality control tests on DNA pool (*see* **Note 5**).

3.4. PCR of SNP From Pooled DNA

1. Allow the DNA pool to come to room temperature and mix gently by vortexing.
2. Pipet, in a dedicated PCR setup area, per sample (use a master mix including calculation for water blank and pipetting error and sample replicates) 12.5 μL 2X Ampitaq Gold PCR master mix, 0.8 μL biotinylated universal primer at 5 pmol/μL, 0.2 μL of tagged SNP primer at 5 pmol/μL, 1.0 μL of untagged primer at 5 pmol/μL, 2.5 μL of DNA at 10 ng/μL, 8.0 μL sterile water: TV 25 μL/sample. Use a heated-lid cycler and use 60 cycles (*see* **Note 6**).

3.5. PCR Plate Setup

1. Use 96-well chimney-top PCR plates. These plates have taller tops for the wells and aid in the PCR template preparation by preventing spillage in downstream manipulations. Set up the plates with eight replicates for each pool (*see* **Note 7**). In addition, for each SNP assay, set up the plate with individual DNA samples representing both homozygotes and the heterozyogote (*see* **Note 8**). Also include the primer controls as described (Pyrosequencing, Biotage AB).

3.6. Template Preparation

1. Immobilize the PCR product and isolate the single-stranded template as described with Dynabeads M-280 strepavidin (Pyrosequencing, Biotage AB) (*see* **Notes 9** and **10**).

3.7. Pyrosequencing

1. Perform as described (Chapter 4) (Pyrosequencing, Biotage AB).

3.8. Allele Quantification

1. Perform analysis of sample wells that contain the individual samples and primer controls for each SNP in the AQ software as described (Pyrosequencing, Biotage AB).

2. Examine the Pyrogram charts of the individual samples. Look for preferential allele amplification, increased A signal and baseline noise (*see* **Note 11**; **Figs. 1** and **2**), and make adjustments.
3. Analyze the sample wells that contain the pooled DNA as described (Pyrosequencing, Biotage AB) with the adjusted factors for SNP assay if required (*see* **Note 12**).
4. Export data to spreadsheet software such as Microsoft Excel™ for further analysis.
5. For association tests, conduct a two-sample test for binomial proportions ***(18)*** to calculate significance of the difference between case and control pools (*see* **Notes 13** and **14**).

4. Notes

1. The SNP-specific primers are designed using the web-based Primer 3 software ***(17)*** (http://frodo.wi.mit.edu/cgi-bin/primer3/primer3_www.cgi). The default software parameters are used and the size of the amplimers range from 100 to 200 bp, although larger templates up to 500 bp and smaller templates down to 80 bp have been used successfully. Avoid repeat nucleotide sequences (homopolymers) adjacent to the SNP: for example, C/AAAAA. Ninety-five percent of the primer sets work with the default conditions (60°C melting temperature) without optimization. Adding the 5' sequence tag to one of the primers does not affect the annealing temperature used in the PCR.
2. Samples have been stored at 4°C or –20°C for more than 10 yr with no degradation or contamination.
3. For the purposes of creating DNA pools, efforts to accurately determine DNA concentrations for each sample are critical, as errors will skew the proportion of each genotype in the pool, especially in pools with smaller numbers of individuals. On the other hand, the larger the number of samples in the pool, the less important the individual quantifications become as random errors in individual samples tend to be minimized in large samples. Optical density measurement (260 nm) should not be used to quantify the DNA samples because substances such as protein and salts will give spurious results.
4. The total volume of the pool can be determined by the number of SNPs to be tested in the overall study. Each SNP assay requires a minimum of 200 ng (25 ng per sample times 8 replicates equals 20 μL of 10 ng/μL) of pooled DNA. To assay approx 2500 SNPs a pool total volume of 50 mL will be required. The volume and concentration of the pool could be adjusted to meet the laboratory's own requirements and availability of DNA.
5. It is prudent to test the accuracy of the making of the DNA pool prior to embarking on a major study. This involves genotyping all the individuals contained in the pool and comparing the allele frequency to that obtained by genotyping the pool itself. Testing SNPs with a wide range of minor allele frequencies (e.g., from 2 to 40%) adds more validity to the quality control. **Table 1** shows the

A

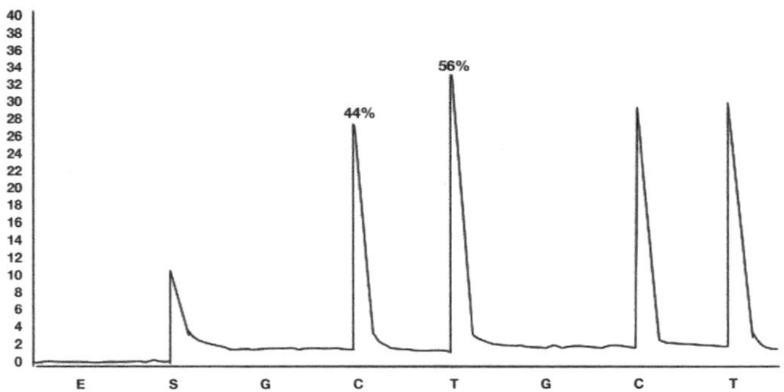

This is a pyrogram for a C/T SNP from a heterozygous individual showing preferential allele amplification of the T allele. The peak heights should be the same amplitude and the ratios 50%C and 50%T. If this is not corrected for before analysis of the pooled data the allele frequency result will be in error.

B

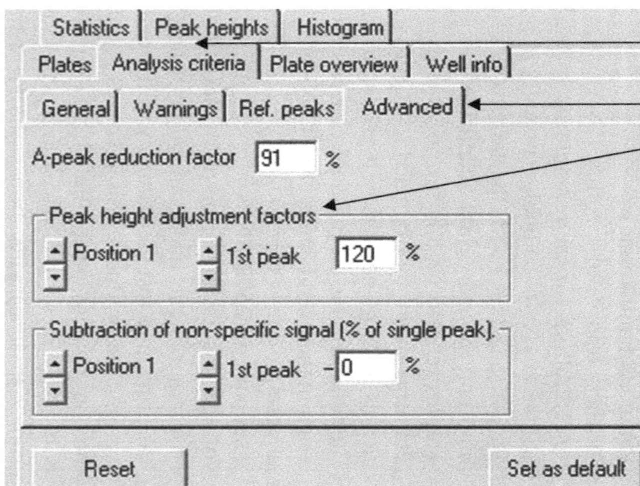

To correct for this, in the AQ software, go to the analysis criteria tab and then go to the advanced tab. The peak height adjustment factors window will be displayed and is adjustable from 80-120%. By changing the factor and reanalyzing the sample the percent C and T can be adjusted until the desired 50%C and 50%T is obtained. In this case, 120% was the final correction factor.

estimation of allele frequencies in pooled DNA samples compared with individual genotyping from one study ***(16)***.

Inaccurate measurement of sample DNA concentrations and noncalibrated pipets are two possible causes of inaccurate pool construction, and this problem would be reflected in a large delta (pool vs individual). What delta is too large? This is a decision that each lab would have to make based on the stringency of their tests. We set the average delta upper limit at 2% for our work with association studies.

6. A 60-cycle PCR incorporates residual primers and dNTPs and makes for a cleaner Pyrosequencing reaction. Use at least 10 ng of genomic DNA in the PCR in order to get enough copies of both alleles included in the amplification reaction to ensure a correct representation of the allele distribution in the pool.
7. A variable number of replicates from 3 to 10 have been tested, and it has been found that 8 replicates most consistently resulted in a standard deviation of 2 or below ***(16)***. The number of replicates for each pool could be modified to meet the stringency of the application.
8. The individual genotypes help define any unusual characteristics about each SNP, such as preferential allele amplification. The primer controls further define the quality of the particular SNP assay, such as baseline noise caused by primer dimerization.
9. For products over 300 bp, a 30-min hybridization at 65°C was preformed with 15-μL strepavidin-coated magnetic beads/sample.
10. Hydra 96-well microdispensing robots were used to aid in the dispensing of the template isolation reagents and to pipet the hybridized PCR products. Robots and multichannel pipets aid in reducing the error and improving the precision of the assays overall.
11. The use of the AQ software is straightforward, however there are some aspects of SNP assays that are important to keep in mind when evaluating the data. The three major concerns are preferential allele amplification, increased signal of the A nucleotide relative to the G, T, and C nucleotides, and baseline noise. In most cases, this can be corrected for by making adjustments to the analysis criteria in the AQ software.

Preferential allele amplification is an artifact of the PCR and results in the amplification of more copies of one allele over the other and is generally owing to the

Fig. 1. *(opposite page)* Correcting for preferential allele amplification. **(A)** This is a pyrogram for a C/T single-nucleotide polymorphism (SNP) from a heterozygous individual showing preferential allele amplification of the T allele. The peak heights should be the same amplitude and the ratios 50% C and 50% T. If this is not corrected for before analysis of the pooled data, the allele frequency result will be in error. **(B)** To correct for this, in the AQ software, go to the analysis criteria tab and then go to the advanced tab. The peak height adjustment factors window will be displayed and is adjustable from 80 to 120%. By changing the factor and reanalyzing the sample the percent C and T can be adjusted until the desired 50% C and 50% T is obtained. In this case, 120% was the final correction factor.

A

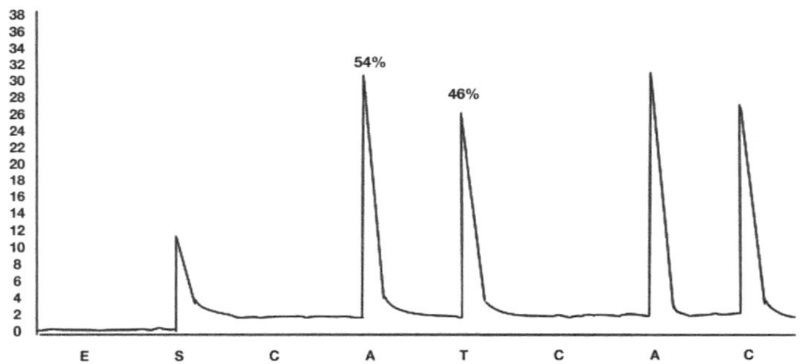

This is a pyrogram for an A/T SNP from a heterozygous individual showing the increased signal of the A nucleotide. The peak heights should be the same amplitude and the ratios 50%A and 50%T. If this is not corrected before analysis of the pooled data the allele frequency result will be in error.

B

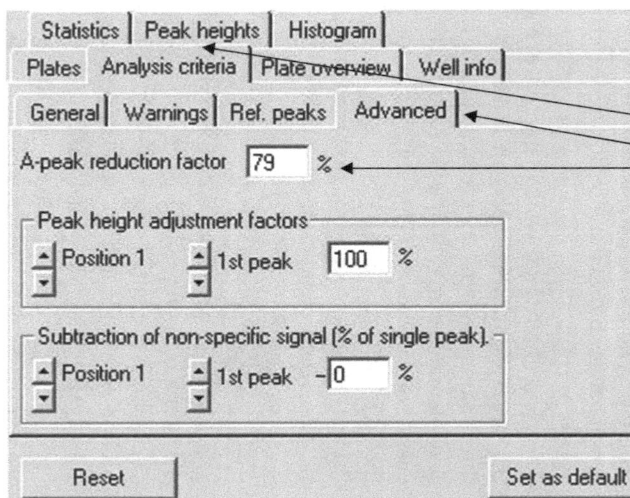

To correct for this, in the AQ software, go to the analysis criteria tab and then go to the advanced tab. The A-peak reduction factor window will be displayed and is adjustable from 70-100%. By changing the factor and reanalyzing the sample the percent A and T can be adjusted until the desired 50%A and 50%T is obtained. In this case, the final correction factor was 79%.

unequal binding properties of the PCR primers. **Figure 1A** shows an example. For this C/T SNP assay (Gs are control bases and should have no signal) the Pyrogram shows preferential amplification of the T allele (C peak height 30 [44% C] and T peak height 36 [56% T]) in a heterozygous individual. The signal for both alleles should be equal in a heterozygous individual. To rule out this being a problem with baseline noise, the homozygous C and homozygous T individuals were examined and found to have no T signal in the homozygous C and no C signal in the homozygous T (data not shown). Additionally, the primer controls for the assay were negative. If baseline noise occurs the assay is redesigned (*see* the last paragraph of **Note 11**). The problem with preferential amplification in this particular assay is that it will lead to over-calculation of the percent T in the pool data. This can be corrected for by the using the "peak height adjustment factor" feature, which can be found under the analysis criteria—advanced tabs in the AQ software **Fig. 1B**. By adjusting the value of the first peak to 120% from the 100% default value and reanalyzing the sample in question, the percent C and T nucleotides were corrected to 50 and 50%. This does not affect the actual data (the T allele in the Pyrogram is still elevated); however, the analyzed data is now corrected and the parameter, once saved, will be applied to all samples for that particular SNP assay. The value can be adjusted from 80 to 120% as necessary. Increased A signal is a problem inherent with the Pyrosequencing reagent chemistry in that the peaks resulting from incorporation of dATPalphaS are slightly higher than other peaks. **Figure 2A** shows an example. For this A/T SNP assay (Cs are control bases and should have no signal), the Pyrogram shows that the amplitude of the A peak is 32 (54% A) and the T peak is 27 (46%) for this heterozygous individual. The peak heights and ratios should be equal in a heterozygous individual. Again, the presence of baseline noise needs to be examined. In the homozygous A and T individuals no baseline T or A signal were detected, respectively, and the primer controls were negative (data not shown). If baseline noise occurs the assay is redesigned (*see* below). Going to the AQ mode and analysis criteria and advanced tab the A-peak reduction factor **Fig. 2B** can be adjusted downward and the sample reanalyzed until the A/T ratios are 50 and 50%, respectively. The default value for this factor is 86% and it can be adjusted from 70 to 100% as necessary. In this case, an A reduction value of 79% was finally used.

Fig. 2. *(opposite page)* Correcting for increased A nucleotide signal. **(A)** This is a pyrogram for a A/T single-nucleotide polymorphism (SNP) from a heterozygous individual showing the increased signal for the A nucleotide. The peak heights should be the same amplitude and the ratios 50% C and 50% T. If this is not corrected for before analysis of the pooled data, the allele frequency result will be in error. **(B)** To correct for this, in the AQ software, go to the analysis criteria tab and then go to the advanced tab. The peak height adjustment factors window will be displayed and is adjustable from 70 to 100%. By changing the factor and reanalyzing the sample the percent C and T can be adjusted until the desired 50% C and 50% T is obtained. In this case, was the final correction factor is 79%.

Table 1
DNA Pool Quality Control

SNP	Pools	MAF in pool	MAF in individuals	Delta (pool vs individual)	Primers
unpublished	A	1.4 + 0.6% T	2.8% T	1.40%	AGCGCTGCTCCGGTTCATAGATTGCCTGTGAAAACGAGAAGAA-rev
size 239bp	B	2.5 + 0.2% T	1.4% T	1.10%	GAGATCCAGATCCGACCACT-fwd
GGC/TTATG					CCTTGCGGGTACCTGG-sequencing fwd
rs285186	A	7.4 + 0.3% G	6.2% G	1.20%	AGCGCTGCTCCGGTTCATAGATTCCTTGCCATGGCCCTGAGAG-rev
size 428bp	B	12.9 + 0.5% G	11.7% G	1.20%	GCCCGCGATGCCTGCTACC-fwd
CTA/GTTTCC					GCTGGTGTGGGAAAGT-sequencing fwd
unpublished	A	17.8 + 0.4% C	18.2% C	0.40%	AGCGCTGCTCCGGTTCATAGATTCGACTGGCTGAA ACAGAATG-rev
size 176bp	B	14.6 + 0.8 % C	15.4% C	0.80%	CCATTGGCTGCTTTTGAA-fwd
C/TATGCGGG					CATTCCACGTTCTTAACT-sequencing fwd
rs717247	A	41.9 + 0.7% G	39.8% G	1.10%	AGCGCTGCTCCGGTTCATAGATTTTATTCCCCTAAAGAGAGGTCA-fwd
size 352bp	B	36.8 + 0.5% G	34.8% G	2.00%	GACAATGCTTAGGTGCTTTCATA-rev
TAA/GGCTTC					GCTTATAAAAAATATTA-sequencing rev

Four SNP assays with a wide range of minor allele frequencies (MAFs) genotyped in DNA pools and in the individuals that are contained in the pools. These single-nucleotide polymorphism (SNP) assays require no adjustments (*see* **Note 11**). MAFs in the pools are the averages of eight replicates. Pools A and B contain 150 Caucasian individuals each. The MAFs for these SNPs may be different in other ethnic groups. The average delta for Pool A is 1.02% and for Pool B is 1.28%. Pools with average deltas greater than 2% are remade.

In some cases, SNP assays can have baseline noise resulting from certain aspects of the Pyrosequencing reaction: self-annealing of the sequencing primer, self-annealing template, or mispriming of the primer in the template. This irregularity is not always reproducible. In our experience the easiest way to circumvent these problems is to redesign the assay with new primers, and the most reliable fix is to design the assay so that the SNP-sequencing primer comes from the other direction. The AQ software has an adjustment factor for baseline noise, but we prefer the redesign assay in these cases. Inevitably, some SNP assays will not work because of noise, and this occurs approx 4% of the time.

12. A time-saving feature of the AQ software is that it will calculate the minimum and maximum allele frequencies and standard deviation for multiple samples simply by pressing the control key and selecting the samples of interest.
13. To assess whether SNP allele frequencies differ between cases and controls the measurement error, the number of individuals tested and the frequency of the SNP are all factors ***(16)***.
14. A limitation of measuring SNP allele frequencies in pools is that one cannot estimate the Hardy–Weinberg equilibrium or construct haplotypes. To obviate this limitation, once we observe allele frequencies that appear to differ between cases and controls, we genotype additional nearby SNPs and then genotype individuals for each SNP and construct haplotypes.

References

1. Ott, J. (1991) *Analysis of Human Genetic Linkage, 2nd ed.*, Johns Hopkins University Press, Baltimore, MD.
2. Hirschhorn, J. N. and Daly, M. J. (2005) Genome-wide association studies for common diseases and complex traits. *Nat. Rev. Genet.* **6,** 95–108.
3. Risch, N. and Merilangas, K. (1996) The future of genetic studies of complex human diseases. *Science* **273,** 1516–1517.
4. Altshuler, D., Hirschhorn, J. N., Klannemark, M., et al. (2000) The common PPAR gamma Pro12Ala polymorphism is associated with decreased risk of type 2 diabetes. *Nat. Genet.* **26,** 76–80.
5. Hugot, J. P., Chamaillard, M., Zouali, H., et al. (2001) Association of NOD2 leucine-rich repeat variants with susceptibility to Crohn's disease. *Nature* **411,** 599–603.
6. Van Eerdewegh, P., Little, R. D., Dupuis, J., et al. (2002) Association of the ADAM33 gene with asthma and bronchial hyperresponsiveness. *Nature* **418,** 426–430.
7. Stefansson, H., Sigurdsson, E., Steinthorsdottir, V., et al. (2002) Neuregulin 1 and susceptibility to schizophrenia. *Am. J. Hum. Genet.* **71,** 877–892.
8. Thomas, D., Haile, R., and Duggan, D. (2005) Recent developments in genomewide association scans: a workshop summary and review. *Am. J. Hum. Genet.* **77,** 337–345.
9. Silander, K., Mohlke, K. L., Scott, L. J., et al. (2004) Genetic variation near the hepatocyte nuclear factor-4 alpha gene predicts susceptibility to type 2 diabetes. *Diabetes* **53,** 1141–1149.

10. Shiffman, D., Ellis, S. G., Rowland, C. M., et al. (2005) Identification of four gene variants associated with myocardial infarction. *Am. J. Hum. Genet.* **77,** 596–605.
11. Deutsch, S., Choudhury, U., Merla, G., Howald, C., Sylvan, A., and Antonarakis, S. E. (2004) Detection of aneuploidies by paralogous sequence quantification. *J. Med. Genet.* **41,** 908–915.
12. Yang, A. S., Estecio, M. R., Doshi, K., Kondo, Y., Tajara, E. H., and Issa, J. P. (2004) A simple method for estimating global DNA methylation using bisulfite PCR of repetitive DNA elements. *Nucleic Acids Res.* **18,** e38.
13. Sun, A., Ge, J., Siffert, W., and Frey, U. H. (2005) Quantification of allele-specific G-protein beta3 subunit mRNA transcripts in different human cells and tissues by Pyrosequencing. *Eur. J. Hum. Genet.* **13,** 361–369.
14. Lavebratt, C., Sengul, S., Jansson, M., and Schalling, M. (2004) Pyrosequencing-based SNP allele frequency estimation in DNA pools. *Hum. Mutat.* **23,** 92–97.
15. Gruber, J. D., Colligan, P. B., and Wolford, J. K. (2002) Estimation of single nucleotide polymorphism allele frequency in DNA pools by using Pyrosequencing. *Hum. Genet.* **110,** 395–401.
16. Wasson, J., Skolnick, G., Love-Gregory, L., and Permutt, M. A. (2002) Assessing allele frequencies of single nucleotide polymorphisms in DNA pools by Pyrosequencing technology. *Biotechniques* **32,** 1144–1150.
17. Rozen, S. and Skaletsky, H. J. (2000) Primer3 on the WWW for general users and for biologist programmers. In: *Bioinformatics Methods and Protocols,* (Krawetz, S. and Misener, S., eds.), Humana Press, Totowa, NJ, pp. 365–386.
18. Rosner, B. (2000) Two sample test for binomial proportions. In: *Fundamentals of Biostatistics,* Duxbury Press, Belmont, CA, pp. 356–371.

7

Multiplex Pyrosequencing® for DNA Variation Analysis

Pritesh Patel, Yih-Horng Shiao, and Paolo Fortina

Summary

Pyrosequencing® technique has been widely used to perform both single-nucleotide polymorphism detection and quantitative DNA methylation analysis. Simplex Pyrosequencing is sufficient to interrogate more than one polymorphic site if these gene variants are within the reach of the sequencing reaction. For polymorphisms far apart from each other or located on different genes, multiple simplex analyses are required. To reduce the number of simplex reactions, multiplex Pyrosequencing becomes a useful alternative method. The multiplex reaction is performed in the presence of single or multiple templates with several sequencing primers. Factors such as primer selection for the PCR and Pyrosequencing reaction, generation of optimal nucleotide dispensation order, use of internal and external controls, preparation of instrumentation, and Pyrogram® interpretation are essential to the success of the multiplexing. In this chapter, the mouse *45S rRNA* gene is used to present two general multiplex Pyrosequencing protocols for determining DNA methylation and allele frequency in the spacer promoter region of this gene.

Key Words: Pyrosequencing®; multiplex; DNA methylation; allele frequency.

1. Introduction

Pyrosequencing® provides an accurate and flexible platform for analyzing DNA sequence variations and has been used in pharmacogenetics, micro-organism detection, clinical molecular research, and DNA methylation analysis *(1–7)*. It is based on the sequencing-by-synthesis principle and relies on real time, indirect, luminometric detection of pyrophophate *(8–13)*. In the Pyrosequencing reaction a sequencing primer anneals to a single-stranded PCR product and extends stepwise in the presence of four enzymes. The released pyrophosphate upon nucleotide incorporation by DNA polymerase is converted to detectable light by ATP sulfurylase and luciferase, which is proportional to

From: *Methods in Molecular Biology, vol. 373: Pyrosequencing® Protocols*
Edited by: S. Marsh © Humana Press Inc., Totowa, NJ

the number of incorporated nucleotides. Any unincorporated nucleotides are degraded by apyrase, hence making the extension reaction ready for the next nucleotide addition. The reaction occurs at constant temperature inside an automated sequencing instrument that delivers precise and ordered nucleotide dispensations. Signals are captured by a CCD camera and graphed in real time in a diagram called Pyrogram®.

In a typical single-nucleotide polymorphism (SNP) detection or DNA methylation analysis, sequence reads of 5–15 bases are generated around and including the polymorphic site(s). The nonpolymorphic peaks are constant and used as internal references to accept or to reject the peak ratio at the polymorphic site(s). During analysis, software compares the peak pattern from the actual run with theoretically predicted Pyrogram patterns for accurate detection of the polymorphic site. Because the signals are quantitative, Pyrosequencing is being increasingly used in applications such as DNA methylation analysis ***(14,15)***.

Thus far, studies on Pyrosequencing have mostly reported its simplex assay capability; however, Pyrosequencing also offers a straightforward approach for designing and performing multiplex assays ***(16–18)***. The essential step in multiplex assay design is to find an optimal dispensation order capable of discriminating multiple nucleotide incorporation patterns in the same run. The feasibility of multiplexing can be first examined by the Pyrosequencing-accompanied software, which provides different combinations of dispensation orders, allowing the operator to select the optimal multiplex assay.

Irrespective of the degree of multiplexing, the Pyrosequencing procedure is generally analogous to simplex Pyrosequencing: first the regions containing DNA polymorphism(s) of interest are PCR amplified, then the Pyrosequencing reaction is performed to detect the polymorphism(s). The major difference is that multiplex Pyrosequencing can include several PCR amplicons, either from multiplex PCR or pooling of simplex PCR products, especially useful for multiple polymorphisms that are farther apart than the upper recommended limit for the Pyrosequencing amplicon (about 300 bp).

Selection of primers for Pyrosequencing assays can be easily accomplished by Pyrosequencing-accompanied software. PCR and Pyrosequencing primer candidates that meet Pyrosequencing analysis criteria (*see* **Note 1**) are automatically listed with scores representing the specificity of each Pyrosequencing reaction. The assay design software computes sequencing primers that are positioned immediately or a few bases before the polymorphism. The sequence of 5–10 bases including the polymorphic site, called the "sequence to analyze," is determined for each template and subsequently entered into the Pyrosequencing-accompanied software to generate the dispensation orders. An optimal dispensation order is essential for a reliable multiplex assay. In this chapter, the utility of multiplex Pyrosequencing is exemplified by providing two assays for determining both the DNA methylation and allele frequency in the mouse *45S ribosomal RNA* (*rRNA*) gene.

2. Materials

2.1. PCR for Pyrosequencing

1. Thermocycler.
2. PCR kit: *Taq* DNA polymerase, PCR buffer, and $MgCl_2$.
3. dNTPs.
4. PCR primers: biotinylated oligonucleotides need to be HPLC purified. The region of differential DNA methylation in the *45S rRNA* gene, containing sites 674 and 676, were amplified using bisulfite-modified template and primers: 5'-GGAAGTGTTTGTGGTGAGG-3' (forward) and 5'-biotin-CACCAACCCTAACATTTTTCC-3' (reverse). For the region containing the 45S-SNP polymorphic site, the primers: 5'-biotin-GCGCGTGAGCGATCTGTA-3' (forward) and 5'-CTGGTCGCCTCACCACAG-3' (reverse) were used.
5. 10% (w/v) Polyacrylamide gel.
6. Polyacrylamide gel electrophoresis equipment.

2.2. Pyrosequencing Analysis

1. Pyrosequencing instrument (Biotage, Westborough, MA): PSQ HS 96A, including PSQ HS 96A 1.2 software and Assay Design Software 1.0, capillary dispensation tips (CDT), reagent dispensation tips (RDT), and CDT-RDT dispensation cartridge.
2. Vacuum prep workstation (Biotage): includes vacuum filtration device with 96-filter probes (Vacuum Prep Tool), vacuum pump, liquid waste container withstanding high vacuum force, and sample prep troughs for Vacuum Prep Tool (plastic, 200 mL).
3. Heating block to hold proprietary PSQ 96-well microtiter plate (Biotage).
4. Plate shaker (1400 rpm): Thermomix R (Eppendorf, Westbury, NY).
5. 96-Well sample prep plate (Sorensen, Salt Lake City, UT).
6. PSQ 96-well microtiter plate (for annealing and reaction) (Biotage).
7. Strip caps or sealing tape.
8. DNA template: PCR product with one biotinylated strand.
9. PyroGold CDT reagents kit (Biotage): enzymes, substrate, and nucleotides included in separate vials. Store PyroGold reagents according to instructions from Biotage.
10. Streptavidin-sepharose HP beads (Amersham Biosciences, Piscataway, NJ).
11. Pyrosequencing primer of high quality (final 0.4 μM): sequencing primer for the 674, 676, and 45S-SNP sites are as follows: 5'-GGAAGTGTTTGTGGTGAGG-3' (674 site), 5'-GATTAGGGTGATAGGAG-3' (676 site), 5-ctcgatcaaccatacaaa (45S-SNP site).
12. High-purity water, 18.2 MΩ-cm or equivalent (Millipore, Bedford, MA).
13. 70% (v/v) Ethanol solution.
14. 0.2 *M* NaOH denaturation solution.

15. Washing buffer: 10 m*M* Tris-acetate, pH 7.6.
16. Binding buffer: 10 m*M* Tris-HCl, 2 *M* NaCl, 1 m*M* EDTA, and 0.1% (v/v) Tween-20, pH 7.6.
17. Annealing buffer: 20 m*M* Tris-acetate and 2 m*M* Mg-acetate, pH 7.6.
18. TE buffer: 10 m*M* Tris-HC1 and 1 m*M* EDTA, pH 8.0.

3. Methods

Here, we describe the process of designing and performing multiplex Pyrosequencing assays with two duplex assays for analyzing differential DNA methylation and a polymorphism in the spacer promoter region of the mouse *45S rRNA* gene.

The first duplex assay, multiplex 674/676, interrogates the methylation status of two CpG sites, named 674 and 676, residing on the same amplicon. Although multiplex assays cannot be quantified with the present software, this example of DNA methylation analysis shows the great potential of multiplex Pyrosequencing. Side-by-side comparisons with simplex assay for each CpG site were performed to demonstrate the accuracy of the multiplex assay. The second assay, multiplex 45SNP/676, examines the 45S-SNP polymorphism and the 676 CpG site simultaneously. These two sites reside on separate amplicons, thus providing an example of the multiplex Pyrosequencing using pooled amplicons.

The methods detail the following: (1) multiplex assay design (for selection of PCR and sequencing primers, and optimal dispensation order); (2) PCR template generation; (3) Pyrosequencing sequential steps (extraction of biotinylated strand, annealing of sequencing primer, and Pyrosequencing reaction using the Pyrosequencing instrument); and (4) Pyrogram interpretation. This protocol confirms the multiplex results with simplex Pyrosequencing and can be adapted to other multiplex Pyrosequencing assays.

3.1. Multiplex Assay Design Scheme

The essential step in multiplex assay design is to find an optimal nucleotide dispensation order capable of discriminating multiple nucleotide incorporation patterns in the same run. Specifically, the peaks from each polymorphic and at least one nonpolymorphic sites of the same primer should not overlap with the peaks from the other extension reactions ***(19)***. Other peaks in the multiplex incorporation pattern may be composite peaks. This requirement, however, may not allow certain "sequences to analyze" to be combined for optimal design. **Figure 1** shows the optimal dispensation orders and theoretical Pyrogram charts obtained for the combination of 674/676 and 45SNP/676, but not 45SNP/674. The general scheme of finding an optimal dispensation order is essentially iterative as outlined in **Fig. 2.**

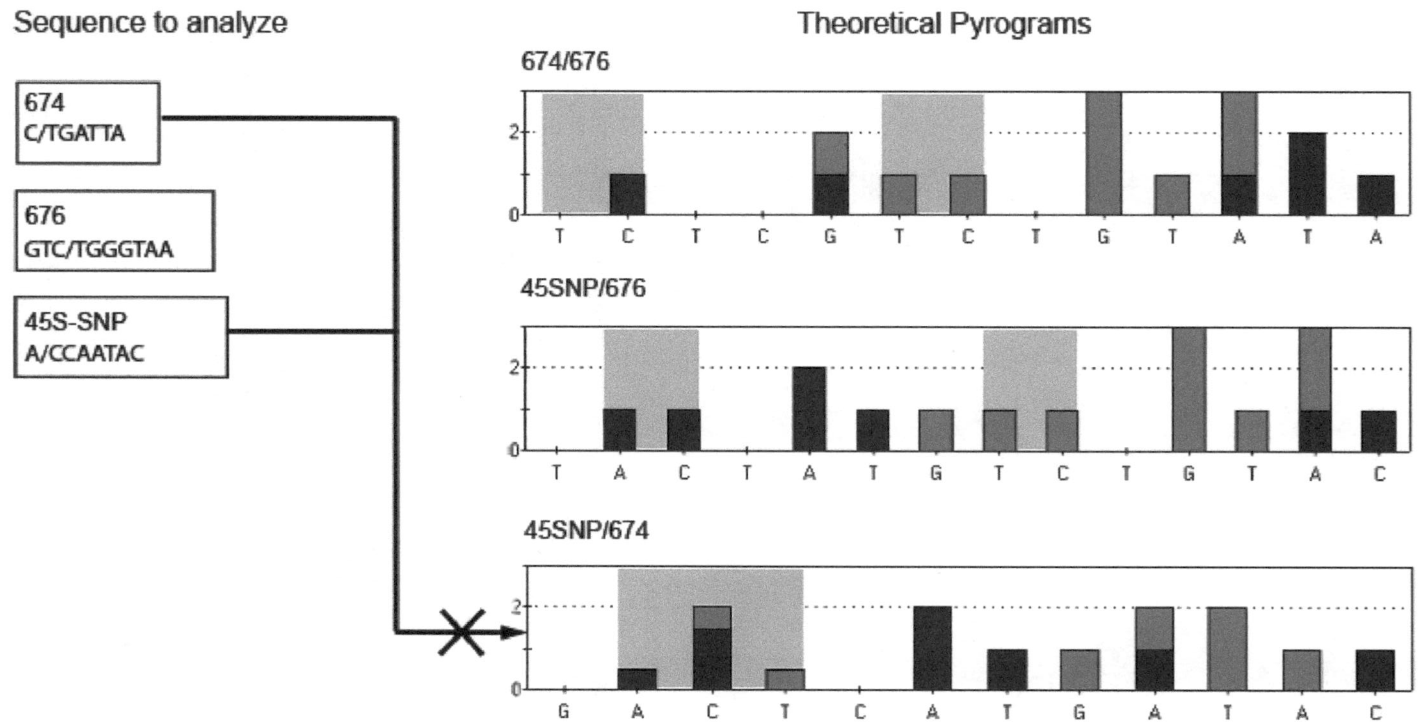

Fig. 1. Compatibility of extension reactions for multiplexing is shown. The "sequence to analyze" and theoretical multiplexing pattern are determined by assay design software 1.0 and PSQ HS 96A 1.2 software, respectively. Black and gray bars represent expected contribution from each of the primer reactions. Areas of shading represent polymorphic sites to be interrogated. The "X" sign indicates the incompatibility of two templates for multiplexing as the theoretical pyrograms show overlapping bars at the polymorphic site.

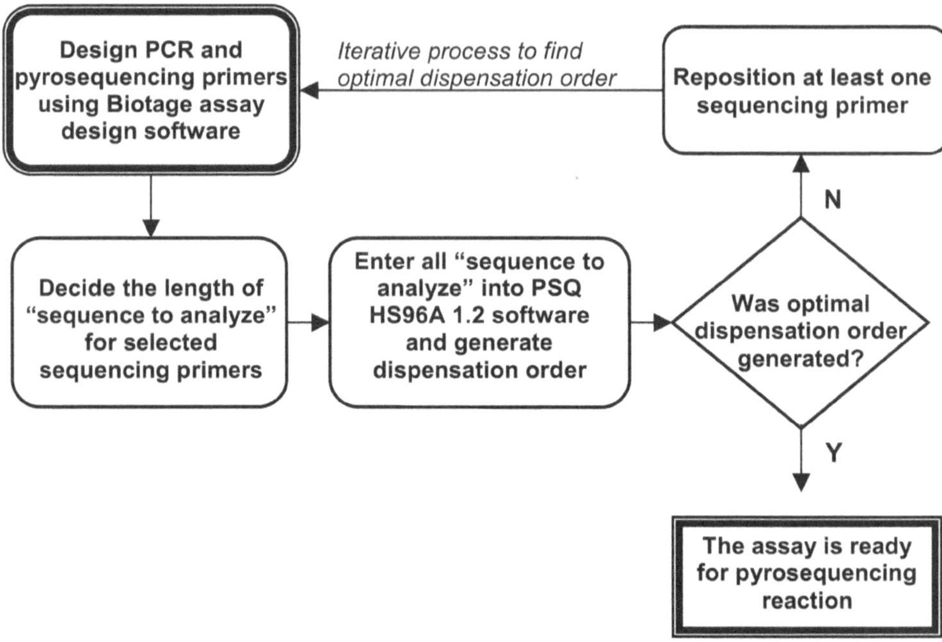

Fig. 2. Multiplex Pyrosequencing® assay design process.

3.1.1. Multiplex 674/676 and Multiplex 45SNP/676 Assay Design

1. Obtain the DNA sequence of the spacer promoter region of the mouse *45S rRNA* gene (GenBank Accession X82564).
2. Enter into the Assay Design 1.0 software, expected sequences after bisulfite modification with a C/T variant at the CpG sites for the DNA methylation assay for the 674 and 676 regions and the wild-type sequence with A/C polymorphism for the 45S-SNP region.
3. The PCR and Pyrosequencing primers with high scores from the Assay Design software are described in **Subheadings 2.1.** and **2.2.** The size of the amplicon covering both 674 and 676 differential DNA methylation sites is 174 bp and the amplicon for the 45S-SNP region is 142 bp. Optimal dispensations and theoretical peak patterns for "sequence to analyze" were obtained from PSQ HS 96A 1.2 software after providing the selected sequencing primers (*see* **Fig. 1**). Peaks are shown as black and gray bars to differentiate signal from different primer extensions.

3.2. PCR for Pyrosequencing (see Note 2)

1. Amplify the region containing sites 674 and 676 in *45S rRNA* gene spacer promotor from 1/10 to 1/5 of bisulfite-modified DNA (from initial 250–1000 ng genomic DNA) for DNA methylation analysis, and the 45S-SNP site region from

unmodified genomic DNA (about 50 ng). Carry out the PCR reaction in a 100-μL reaction mixture containing 20 m*M* Tris-HCl (pH 8.0), 50 m*M* KCl, 1.5 m*M* $MgCl_2$, 0.2 m*M* dNTPs, 2.5 U platinum *Taq* polymerase, and 0.2 μ*M* primers (*see* **Subheading 2.1.** for PCR primer sequences for the 674/676 and the 45S-SNP regions).
2. Perform initial denaturation at 94°C for 2 min, then 35 cycles of denaturation (94°C, 30 s), annealing (50°C for 674/676 and 60°C for 45S-SNP, 30 s), and extension (72°C, 1 min). Increase the last extension to 5 min.
3. Confirm specific amplification and sufficient PCR yield by 10% (w/v) polyacrylamide gel electrophoresis and ethidium bromide (0.5 μg/mL) fluorescence staining. It is important that the amplification is free of extraneous products and that the yield is sufficient (instrument model dependent) to achieve robust Pyrosequencing results.

3.3. Pyrosequencing Assay

Prior to carrying out the following steps for a large set of samples, ensure that the assay has been validated with controls (*see* **Note 3**) and that the instrument and reagents have been properly prepared (*see* **Notes 4** and **5**).

3.3.1. Extraction of Biotinylated Single-Stranded DNA

1. Allow all solutions to reach room temperature before starting.
2. Add 5 μL PCR product, 2 μL homogenous streptavidin-sepharose beads, 40 μL binding buffer, and 35 μL high-purity water into each 96-well sample prep plate well. (**Note**: for pooling of two amplicons add 5 μL of each PCR product and reduce water to 30 μL.)
3. Cover the plate with adhesive film and incubate the mixture on a plate shaker for at least 5 min, agitating at 1400 rpm at room temperature. Proceed to **step 5** immediately after shaking is finished because sepharose beads sediment quickly (shaking can continue for a little longer while preparing materials at **step 4**).
4. Although the PCR product and beads are mixing, prepare the PSQ 96-well microtiter plate containing 0.4 μ*M* of each sequencing primer in a total annealing buffer volume of 12 μL in corresponding wells. In addition, set up the vacuum prep workstation for strand separation according to the "Sample Preparation Guidelines for PSQ HS 96A Systems" from Biotage.
5. Bring the 96-well sample prep plate from **step 3** to the vacuum prep workstation. Prime the Vacuum Prep Tool by lowering it into a plastic trough containing high-purity water for 30 s and allow water to flush through the tool.
6. Capture the sepharose streptavidin-bound PCR products by immersing the vacuum prep tool into the 96-well sample prep plate ***(20)***. The vacuum pressure should be set to a minimum of 300 mmHg. The liquid is flushed through the filter while the bead-bound PCR products are retained on the filter probes.
7. Transfer the vacuum prep tool to a trough containing 70 % (v/v) ethanol for 5 s.
8. Denature and remove the nonbiotinylated DNA strands by transferring the vacuum prep tool to a trough of denaturation solution for 5 s.

9. Wash bead-bound biotinylated strands by moving the vacuum prep tool to a trough of washing buffer for 7 s.
10. Raise the vacuum prep tool vertically for a few seconds to drain the excess washing buffer and then return it to horizontal position before turning off the vacuum.
11. Release the beads from the filter by lowering the vacuum prep tool and immersing the filter probes into the PSQ 96-well microtiter plate (optical) containing sequencing primers and shake the tool with filter probes resting at the bottom of the wells.

3.3.2. Annealing of Sequencing Primers and Execution of Pyrosequencing Reaction

1. Incubate the microtiter plate at 90°C for 2 min in the heating block, then cool at 45°C for 10 min to facilitate annealing of the Pyrosequencing primer(s) to single-stranded template(s) (*see* **Note 6**).
2. Transfer the plate to the PSQ HS 96A Pyrosequencing instrument immediately. Do not allow the plate to cool off too fast.
3. Place the annealing/reaction plate into the PSQ HS 96A Pyrosequencing instrument process chamber and directly execute the run on the ready instrument with "sequence to analyze" and sample information being entered.

3.4. Pyrogram Interpretation

3.4.1. Multiplex 674/676

Pyrogram charts from 674/676 multiplex and simplex assays are illustrated in the top panel of **Fig. 3**. The peaks in the highlighted area represent the C/T polymorphic sites. Percentage of C, indicative of methylated cytosine, is reported in the simplex reaction Pyrogram. Several dispensations that are not in the original sequence are automatically added generally after the substrate peak (S in the Pyrogram) and polymorphic peaks by the software to function as internal negative controls. Negative controls should have no significant background signal. Remaining nonvariable peaks in each extension reaction act as positive controls and reference peaks. For multiplex 674/676, the Pyrogram is similar to the composite pattern of the 674 and 676 simplex reactions. The ratios of C/T at sites 674 and 676 in simplex assays match up with signals in the multiplex reaction. The actual multiplex Pyrogram corresponds very well with the theoretically predicted pattern (*see* **Fig. 1**). If the nonvariable peaks do not conform to the predicted pattern, the Pyroseqeuncing software flags the polymorphic peaks with different colors signaling either "check" or "failed" for that reaction.

3.4.2. Multiplex 45S-SNP/676

Figure 3, bottom panel, shows the Pyrogram charts from multiplex and simplex reactions of the 45S-SNP and the 676 regions. The peak heights at both

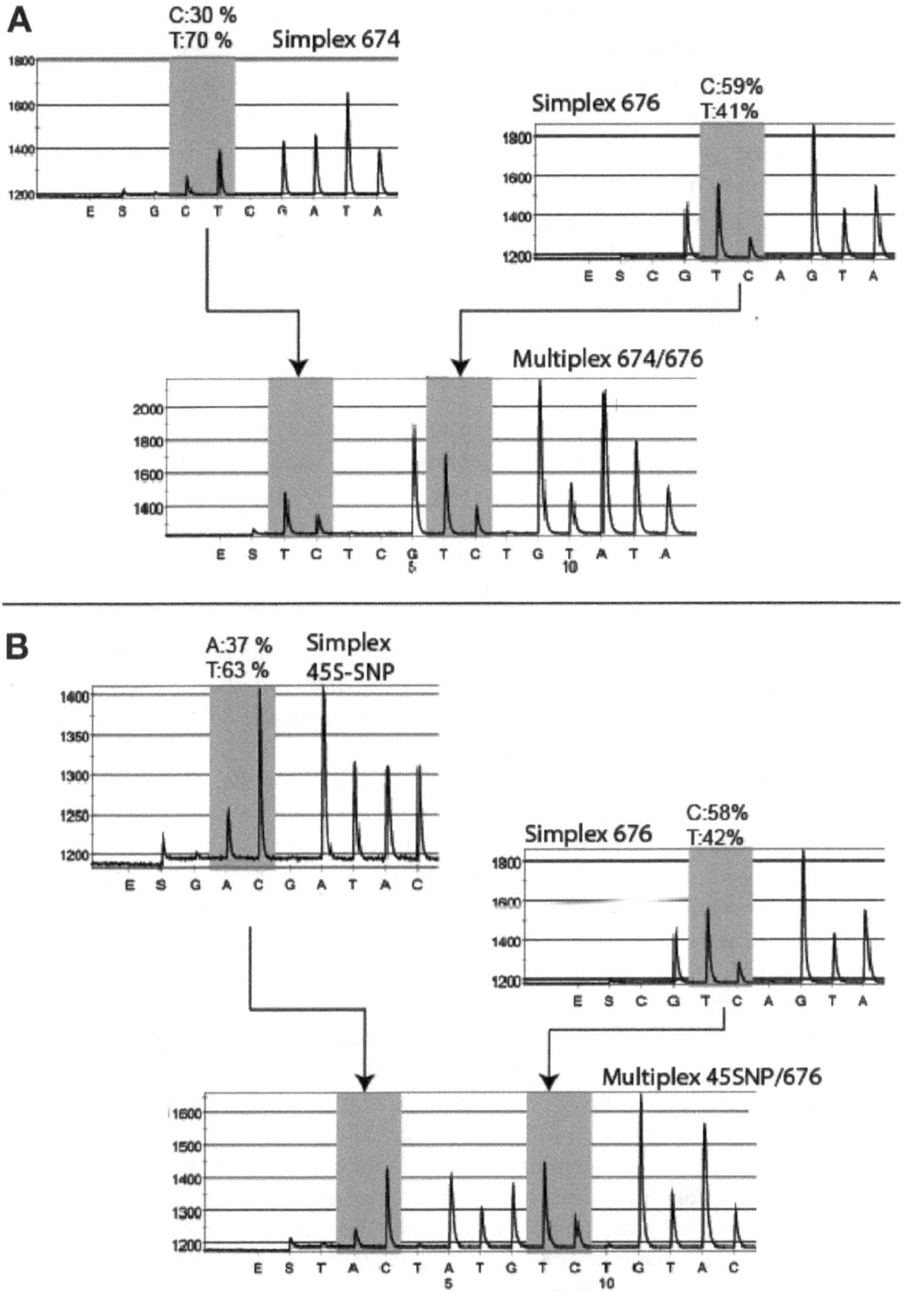

Fig. 3. Multiplex assay Pyrogram® charts compared to simplex assay Pyrogram charts. **(A)** Multiplex 674/676 compared to Simplex 674 and 676 Pyrogram charts. **(B)** Multiplex 45SNP/676 compared to Simplex 45SNP and 676 Pyrogram charts. Arrows show the corresponding variation sites.

polymorphic sites from multiplexing agree perfectly with the signals from separate simplex reactions. Because it is extremely difficult to mix equal molar ratio of 45S-SNP and 676 PCR products, the peaks from two different extension reactions are expected to vary in height (*see* **Note 7**). To overcome the unequal amount of multiple templates for Pyrosequencing, the PSQ HS 96A software has been designed to take a peak from the same primer extension as a reference to determine the accuracy of the polymorphic peak heights. Therefore, it is important to select a dispensation order showing that polymorphic and at least one nonpolymorphic peak are from the same extension reaction (*see* **Fig. 1**; *see also* **Note 8**).

4. Notes

1. Either the Primer Design software or the latest version, Assay Design 1.0 (Biotage), is very effective in selecting PCR and sequencing primers for Pyrosequencing. Other primer-selecting programs may be used but require complex manipulation to achieve the same type of analysis. The PCR product should be ideally shorter than 300 bp and have a low penalty score in self-annealing and hairpin formation. Sequencing primer and biotinylated PCR primer should be high quality and can be purified by high-pressure liquid chromatography to ensure full-length primers. The above Pyrosequencing designing programs examine and give scores for potential secondary structure, primer length, melting temperature (please note that the Pyrosequencing reaction occurs at 28°C), and homopolymers adjacent to polymorphic site.
2. One of the PCR primers needs to be biotinylated. Keep the primer concentration low to prevent excessive biotinylated primer from competing with PCR products for streptavidin-binding sites. The PCR reaction should be free of nonspecific amplification, and needs to be empirically optimized to obtain sufficient products for Pyrosequencing with minimal cycle number. Overamplification may alter the ratio of polymorphic nucleotides. The amount of PCR products can be estimated simply by comparing with the intensity of DNA molecular weight marker of a known concentration.
3. It is imperative to test some controls first to ensure that the Pyrogram is accurate. Assay specificity and hence genotyping can be compromised if substantial background signal, false peaks, or partial peaks are generated from sources other than the sequences to be analyzed. Such sources include nonspecific binding of the sequencing primer, self-priming of the template, primer-dimer, and hairpin formation of sequencing primer. In multiplex assays, the risk of generating false background signal increases because two or three templates are analyzed simultaneously. Assessment of nonspecific signal contribution to the Pyrogram can be easily achieved by including control reactions. Recommended control reactions for multiplex Pyrosequencing assay are as follow: (1) single DNA template (PCR products) only, without sequencing primer; (2) multiple DNA templates only, without Pyrosequencing primers; (3) single sequencing primer only, without

DNA template; (4) multiple sequencing primers only, without DNA templates; and (5) cross check of the DNA template with the sequencing primer for other template(s), if multiple PCR products are meant to be pooled in Pyrosequencing. These can be performed initially when an assay is being validated and included in the routine analysis.

4. Prerun preparation of the PSQ HS 96A instrument is well described in the manufacturer's manual and needs to be done for each run. Briefly, the steps are: (1) dilute PyroGold reagents (enzyme and substrate solutions) and nucleotides; (2) load CDT and RDT with appropriate amounts of nucleotides and reagents, respectively; and (3) insert the assembled cartridge into the instrument and perform the dispensation test. The dispensation test is a crucial instrumental function that can be accessed from the PSQ 96A HS 1.2 software. The nucleotides need to be diluted 1:1 in TE buffer, pH 8.0. This helps to reduce the surface tension and to allow better nucleotide dispensation.
5. Dust, air bubbles, or the reagents themselves (in particular viscous reagents) can block the tips. It is important to use dust-free gloves and filter-free pipet tips, to work in a dust-free environment and to avoid introducing air bubbles while handling the tips. To ensure that dispensation tips are unblocked, perform dispensation tests prior to running the samples. Centrifugation of nucleotide vials for 10 min at maximum speed helps prevent particles in the reagent from being transferred to the dispensation tips. Avoid leaving the substrate dispensation tips in the instrument for longer than 5 min unless it is running. If needed, a dispensation test can be performed every 5 min as a means of stirring. Because the tips clog so easily, it is imperative to wash the tips immediately after the run according to the manufacturer's instruction.
6. Annealing is typically performed by heating the annealing/reaction plate at 80°C for 2 min, followed by cooling it at room temperature. In the work reported here, the reactions benefit from heating at 90°C for 2 min, followed by 10 min of cooling at 45°C on the heating block and immediately running the plate.
7. If the polymorphisms to be interrogated are located far apart in the gene sequence, PCR of different regions will need to be performed, either as multiple singlex reactions followed by pooling of products or direct multiplex PCR. In either case, equimolar representation of amplicons, although not necessary, can certainly make Pyrogram interpretation easier.
8. Some assays may benefit from manually modifying the computer-generated dispensation order to reduce background and unusual peak(s). These reaction artifacts may arise from excess sequencing primer or unknown nucleotide variant, and can result in the software warning of a "failed" Pyrosequencing reaction. For high background in the internal negative control dispensation or the appearance of one outlier peak, manual modification of dispensation order is suggested.

Acknowledgments

The research was supported in part by the Intramural Research Program of the National Institutes of Health, National Cancer Institute, Center for Cancer

Research (Y-H. S), and in part by the PA Department of Health (P.P and P.F). The authors would like to thank Dr. Saul Surrey for critical reading of the manuscript.

References

1. O' Meara, D., Wilbe, K., Leitner, T., Hejdeman, B., Albert, J., and Lundeberg, J. (2001) Monitoring resistance to human immunodeficiency virus type 1 protease inhibitors by Pyrosequencing. *J. Clin. Microbiol.* **39,** 464–473.
2. Ronaghi, M. and Elahi, E. (2002) Pyrosequencing for microbial typing. *J. Chromatogr. B. Analyt. Technol. Biomed. Life Sci.* **25,** 67–72.
3. Ferraris, A., Rappaport, E., Santacroce, R., et al. (2002) Pyrosequencing for detection of mutations in the Connexin 26 (GJB2) and Mitochondrial 12S RNA (MTRNR1) genes associated with hereditary hearing loss. *Human Mutat.* **20,** 312–320.
4. Berg, L. M., Sanders, R., and Alderborn, A. (2002) Pyrosequencing technology and the need for versatile solutions in molecular clinical research. *Expert Rev. Mol. Diagn.* **2,** 361–369.
5. Colella, S., Shen, L., Baggerly, K. A., Issa, J. P., and Krahe, R. (2003) Sensitive and quantitative universal Pyrosequencing methylation analysis of CpG sites. *Biotechniques* **35,** 146–150.
6. Zackrisson, A. and Lindblom, B. (2003) Identification of CYP2D6 alleles by single nucleotide polymorphism analysis using Pyrosequencing. *Eur. J Clin Pharmacol.* **59,** 521–526.
7. Ahluwalia, R., Freimuth, R., McLeod, H. L., and Marsh S. (2003) Use of Pyrosequencing to detect clinically relevant polymorphisms in dihydropyrimidine dehydrogenase. *Clin. Chem.* **49,** 1661–1664.
8. Ronaghi, M., Uhlen, M., and Nyren, P. (1998) A sequencing method based on real-time pyrophosphate. *Science* **281,** 363–365.
9. Alderborn, A., Kristofferson, A., and Hammerling, U. (2000) Determination of single-nucleotide polymorphisms by real-time pyrophosphate DNA sequencing. *Genome Res.* **10,** 1249–1258.
10. Ahmadian, A., Gharizadeh, B., Gustafsson, A. -C., et al. (2000) Single-nucleotide polymorphism analysis by Pyrosequencing. *Anal. Biochem.* **280,** 103–110.
11. Fakhrai-Rad, H., Pourmand, N., and Ronaghi, M. (2002) Pyrosequencing™: An accurate detection platform for single nucleotide polymorphisms. *Human Mutat.* **19,** 479–485.
12. Ronaghi, M. (2003) Pyrosequencing for SNP genotyping. In: *Single Nucleotide Polymorphisms*, (Kwok, P.-Y., ed.), Humana Press, Totowa, NJ, pp. 189–195.
13. Langaee, T. and Ronaghi, M. (2005) Genetic variation analyses by Pyrosequencing. *Mutat Res.* **573,** 96–102.
14. Dupont J. M., Tost, J., Jammes, H., and Gut, I. G. (2004) De novo quantitative bisulfite sequencing using the Pyrosequencing technology. *Anal. Biochem.* **333,** 119–127.

15. Shiao, Y.-H., Crawford, E. B., Anderson, L. M., Patel, P., and Ko, K. (2005) Allele specific germ cell epimutation in the spacer promoter of the 45S ribosomal RNA gene after Cr(III) exposure. *Toxicol. Appl. Pharmacol.* **205,** 290–296.
16. Pourmand, N., Elahi, E., Davis, R. W., and Ronaghi, M. (2002) Multiplex Pyrosequencing. *Nucleic Acids Res.* **30,** e31.
17. Lötsch, J., Skarke, C., and Geisslinger, G. (2003) Simultaneous screening for the three mutations in the ABCB1 gene. *Genomics* **82,** 503–510
18. Palmieri, O., Toth, S., Ferraris, A., et al. (2003) Card15 genotyping in inflammatory bowel disease patients by multiplex Pyrosequencing. *Clin. Chem.* **49,** 1675–1679.
19. Pyrosequencing Technical note 105. URL: http://www.Pyrosequencing.com.
20. Dunker, J., Larsson, U., Petersson, D., et al. (2003) Parallel DNA template preparation using a vacuum filtration sample transfer device. *Biotechniques* **34,** 862–868.

8

Analysis of Gene-Specific DNA Methylation Patterns by Pyrosequencing® Technology

Jörg Tost and Ivo Glynne Gut

Summary

As the sequence of the human genome is now nearly finished, genome research turns to elucidate gene function and regulation. DNA methylation is of particular importance for gene regulation and is strongly implicated in the pathogenesis of various diseases. The real-time luminometric detection of pyrophosphate release upon nucleotide incorporation in the Pyrosequencing® technology is ideally suited for the simultaneous analysis and quantification of the methylation degree of several CpG positions in close proximity. We developed and improved this analysis to obtain reproducible results for as many as 10 successive CpGs in a single sequencing reaction spanning up to 80 nt. Advantages of the Pyrosequencing technology are the ease of its implementation, the high quality and the quantitative nature of the results, and its ability to identify differentially methylated positions in close proximity, which may be used as DNA methylation markers.

Key Words: Pyrosequencing®; methylation; CpG island.

1. Introduction

DNA methylation is the only genetically programmed DNA modification in mammals. This postreplication modification is almost exclusively found on the five position of cytosines in the context of the dinucleotide sequence CpG and is carried out by methyltransferases, of which so far five have been identified *(1)*. The CpG dinucleotide is underrepresented in the genome and is generally methylated except for relatively CpG-rich clusters of approx 1–2-kb length CpG islands that are found in the promoter region and the first exon of many genes. Methylation patterns may be cell-type and/or tissue-specific and change slowly with age and in response to environmental effects such as diet. Cytosine methylation is essential for mammalian embryogenesis and development, and is implicated in X chromosome inactivation *(2)*. It plays a key role in the maintenance of

From: *Methods in Molecular Biology, vol. 373: Pyrosequencing® Protocols*
Edited by: S. Marsh © Humana Press Inc., Totowa, NJ

genome integrity by transcriptional silencing of repetitive DNA sequences and endogenous transposons. DNA methylation is also critical for imprinting *(3)*, i.e., the asymmetric expression of either the maternal or paternal allele in a parent-of-origin-specific manner in somatic cells of the offspring. Imprinting is essential for normal mammalian development, growth, and behavior and is regulated in a developmental- and tissue-specific manner. Methylation of CpG islands is important for the establishment and maintenance of cell-type-restricted gene expression and is often associated with the inhibition of gene expression, but is not a necessity for gene silencing. Aberrant methylation patterns have been reported in various neurodevelopmental disorders and imprinting anomalies lead to disorders such as Prader–Willi, Angelman, and Beckwith–Wiedemann syndrome. A possible role for epigenetic events in complex, non-Mendelian disorders like type 2 diabetes has been postulated *(4)*. A cumulative series of genetic and epigenetic alterations leading to unregulated cell growth is the foundation of tumorigenesis. Epigenetic changes occur early in the progression process and often precede malignancy *(5)*. Global hypomethylation of the genome is observed and has been suggested to initiate and propagate oncogenesis by inducing chromosome instabilities and transcriptional activation of oncogenes and latent retrotransposons. The overall decrease in DNA methylation is accompanied by a region- and gene-specific hypermethylation of CpG islands. Genes involved in DNA repair, detoxification, cell cycle regulation, and apoptosis are often inappropriately inactivated. It has been suggested that epigenetic lesions in normal tissue set the stage for neoplasia. Methylation patterns can be shared by different types of tumors as well as being tumor type-specific, and the extent of hypo- and hypermethylation often correlates with the malignancy grade and/or the disease stage *(6)*.

For the analysis of DNA methylation sensitive and quantitative methods are needed to detect even subtle changes in the degree of methylation as biological samples often represent a heterogeneous mixture of different cells, especially tumor and nontumor cells from tissue biopsies. Realizing the importance of epigenetic changes in development and disease, a number of techniques for the study of DNA methylation has been developed in recent years *(6–8)*. Most methods are either based on methyl-sensitive enzymatic digestion of genomic DNA or rely on bisulfite conversion. However, most of the current protocols are either labor intensive such as cloning and subsequent sequencing, or provide information only on one or two CpG positions such as MS-SNuPE, the GOOD assay for epigenotyping, or COBRA. DNA methylation analysis by the Pyrosequencing *(9)* technology overcomes these limitations *(10–13)*. A cascade consisting of four enzymes and specific substrates produces a light signal upon nucleotide incorporations and subsequent pyrophosphate release. The light signal is detected via a CCD camera and is converted into a quantitative signal in the Pyrogram®. Pyrosequencing® can be applied to the determination of global DNA methylation content of a sample ***(14–16)*** as well as gene-specific analyses ***(10–13,17)***. For the latter it is ideally suited as an epigenotype-mapping tool as it allows an unbiased quantitative analysis of up to 10 CpG positions in a single sequencing run of up to 80 nt ***(13)***.

2. Materials

2.1. Assay Design

1. Design of PCR primers using MethPrimer ***(18)*** (http://www.ucsf.edu/urogene/methprimer/index1.html).
2. Design of Pyrosequencing primers using the online software (free of charge for Pyrosequencing users; http://techsupport.Pyrosequencing.com).

2.2. Bisulfite Treatment

1. Sheared or restriction enzyme-digested genomic DNA (1 μg in a volume of 20 μL).
2. Sodium bisulfite (Sigma, St. Louis, MO, cat. no. S-9000), prepare freshly a 5 *M* solution.
3. Sodium hydroxide pellets (Aldrich, St. Louis, MO, cat. no. 480878), prepare a 5 *M* stock solution and store in plastic bottles.
4. Hydroquinone (Sigma, cat. no. H-9003).
5. Wizard® DNA clean-up system (Promega, Madison, WI, cat. no A7280).
6. Isopropanol.
7. Ammonium acetate.

2.3. PCR

1. Primers for PCR amplification (Biotez, Buch, Germany).
2. Biotinylated primers for PCR amplification (Biotez).
3. Platinum *Taq* DNA polymerase (Invitrogen Life Technologies, Carlsbad, CA) or HotStar *Taq* DNA polymerase (Qiagen, Hilden, Germany).
4. dNTPs.
5. Thermo-Fast® 96-skirted 96-well plates (ABgene, Epsom, UK, cat. no. AB-0800).
6. Eppendorf 96-Gradient Mastercycler (Eppendorf, Hamburg, Germany).

2.4. Purification and Preparation of Single-Stranded Template

1. Streptavidin-sepharose HP beads (GE Healthcare, Uppsala, Sweden, cat. no. 17-5113-01).
2. Vacuum preparation tool ***(19)*** (Biotage, Uppsala, Sweden) using the corresponding filter probes (cat. no. 60-0180).
3. Troughs (Biotage, cat. no. 60-0182).
4. Binding buffer: 10 m*M* Tris-HCl, 2 *M* NaCl, 1 m*M* EDTA, and 0.1% Tween-20, pH 7.6.
5. Denaturing solution: 0.2 *M* NaOH.
6. Washing buffer: 10 m*M* Tris-acetate, pH 7.6 (adjust with 4 *M* acetic acid).

7. Annealing buffer: 20 m*M* Tris-acetate, 2 m*M* Mg-acetate, pH 7.6 (adjust with 4 *M* acetic acid).
8. Thermomixer or similar (room temperature).
9. Heating device, for example heating plate or thermoblock.
10. Thermoplate for sample preparation (Biotage, cat. no. 60-0092).
11. Primers for Pyrosequencing (Biotez).
12. Plate for Pyrosequencing analysis, PSQ 96-plate low (Biotage, cat. no. 40-0010).

2.5. Pyrosequencing Analysis

1. PSQ 96MA System SNP/SQA (Biotage, cat. no. 60-0140).
2. Cartridge for reagent dispensation, PSQ 96 reagent cartridge (Biotage, cat. no. 40-0022).
3. Kit for a read-length of up to 30 nt dispensations: PyroGold SNP reagent kit 5 × 96 (Biotage, cat. no. 40-0044).
4. Kit for a read-length for more than 30 nt dispensation: PyroGold SQA reagent kit 1 × 96 (Biotage, cat. no. 40-0045).

3. Methods

The methods described next outline (1) the bisulfite treatment to "freeze" the methylation status of the sample to be analyzed (**Fig. 1**), (2) the experimental design for a quantitative gene- or promoter-specific analysis of CpG methylation, (3) the PCR amplification of the target sequence, (4) sample preparation for Pyrosequencing analysis, and (5) analysis in the Pyrosequencing instrument. The procedure is also depicted in **Fig. 2**.

3.1. Bisulfite Treatment of Genomic DNA

The introduction of sodium bisulfite conversion of genomic DNA had great impact on the field of DNA methylation analysis ***(20)***. Bisulfite treatment of genomic DNA samples results in the hydrolytic deamination of nonmethylated cytosines to uracils, whereas methylated cytosines are resistant to conversion (**Fig. 1**). After a PCR the methylation status at a given position is manifested in the ratio C (former methylated cytosine) to T (former nonmethylated cytosine) translating epigenetic information into sequence information and can be analyzed as a virtual C/T polymorphism in the bisulfite-treated DNA. The different steps are outlined in the following paragraphs. Probably the most critical step in the bisulfite conversion is the denaturing of the DNA as only single-stranded DNA is accessible to chemical modification. It should be noted that there are also several commercial kits on the market that might facilitate implementing bisulfite conversion.

3.1.1. Preparation and Denaturing of Genomic DNA

The size of genomic DNA can be reduced by either enzymatic digestion or heat-induced fragmentation (100°C, 5 min). One microgram denatured

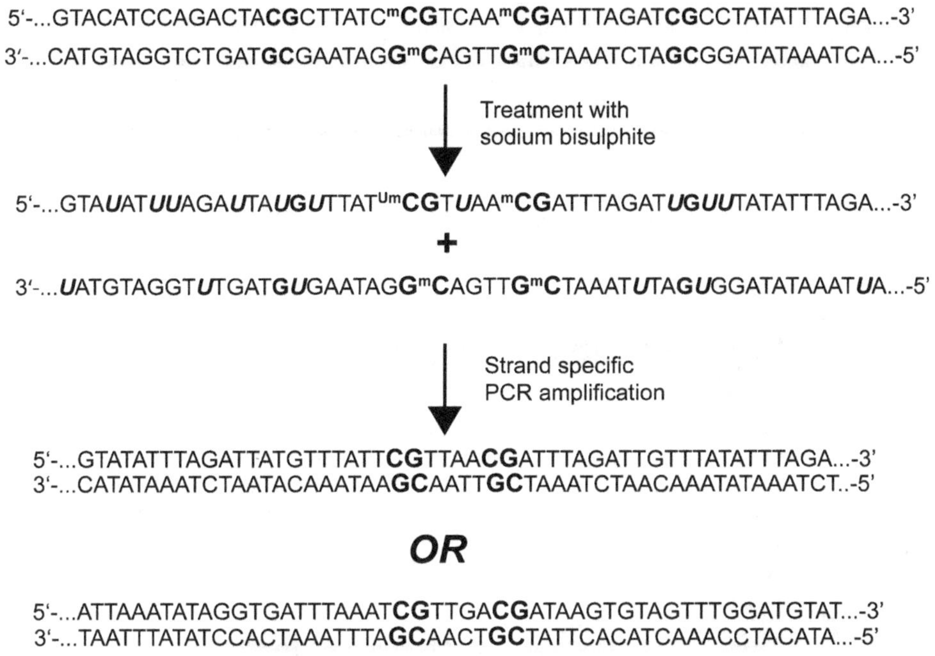

Fig. 1. Outline of the bisulphite treatment. Nonmethylated cytosines are deaminated to uracils that are subsequently replaced by thymines during PCR amplification while methylated cytosines are resistant to conversion under the applied conditions. The former two complementary strands are not symmetrically modified by the bisulfite treatment thereby yielding a C-rich strand and a G-rich strand that are available for amplification and analysis of DNA methylation with the same information content.

genomic DNA is diluted into 48 μL H_2O and incubated with 2 μL of 5 *M* NaOH (final concentration 0.2 *M*) for 15 min at 37°C (*see* **Note 1**).

3.1.2. Sodium Bisulfite Reaction (see **Note 2**)

1. Dissolve 9 mg of hydroquinone in 1 mL H_2O.
2. Add 333 μL of this solution to 2.7 g of sodium bisulfite.
3. Add 400 μL of 5 *M* NaOH.
4. Adjust to a final volume of 5 mL with H_2O (the volume has to be respected otherwise sodium bisulfite will not completely dissolve).
5. Add 550 μL of the bisulfite/hydroquinone solution to the 50 μL of freshly denatured DNA.
6. Incubate for 16 h in a water bath at 50°C in darkness.

3.1.3. Purification, Desulfonation, and Precipitation

1. Add 1 mL of Wizard DNA clean-up resin per sample treated with sodium bisulfite and mix gently.

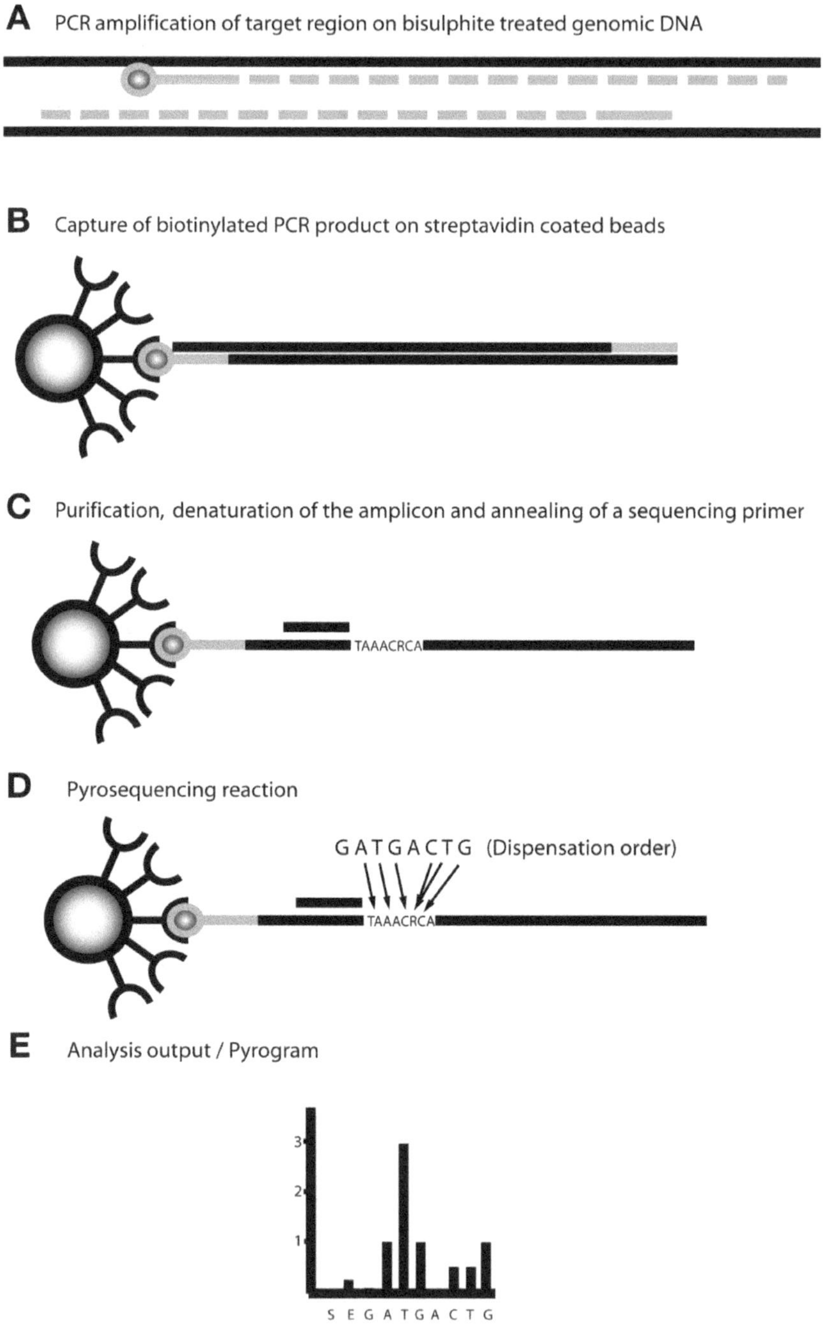

Fig. 2. Outline of the procedure for the analysis of DNA methylation patterns by the Pyrosequencing® technology. The preceding bisulfite conversion is shown in **Fig. 1**.

2. Attach a 2-mL syringe barrel to the Promega minicolumn above a Falcon tube, transfer the sample into the syringe, and apply it slowly onto the column with the help of the piston or a vacuum manifold.
3. Wash the column with 2 mL isopropanol (80%).
4. Transfer the mini-column to a 1.5-mL microcentrifuge tube. Centrifuge the mini-column at 12,000*g* in a microcentrifuge for 20 s to remove any residual isopropanol.
5. Transfer the mini-column to a new microcentrifuge tube and elute the DNA from the resin by adding 50 µL of prewarmed H_2O (70–80°C). After 1 min of incubation, centrifuge the mini-column for 20 s at 12,000*g*.
6. Add 5.5 µL of 3 *M* NaOH (0.3 *M* final) and incubate 5 min at room temperature.
7. Add 5.6 µL of 3 *M* sodium acetate and 200 µL absolute ethanol to precipitate the DNA following standard procedures for DNA precipitation.
8. Dissolve the precipitated DNA in 25 µL H_2O and store at –80°C until use.

3.2. Experimental Design

The design of amplicons including their respective amplification primers on bisulfite-treated DNA is described in **Subheading 3.2.1.**, and the design of the Pyrosequencing primers in **Subheading 3.2.2.** Great attention should be paid to the experimental design as this will crucially influence the successful outcome of the assay. The treatment of genomic DNA with sodium bisulfite greatly reduces the complexity of the genome as the four-letter genetic code is converted to a three-letter "alphabet." However, as the former complementary DNA strands are differentially modified, two populations of DNA molecules are available for primer design as CpG methylation is normally a symmetric modification of the DNA.

3.2.1. Design of Amplification Primers

Primers for amplification can be designed using the MethPrimer software ***(18)***.

1. As the two strands of genomic DNA are no longer complimentary after bisulfite treatment, both strands can be analyzed for possible amplification products. MethPrimer takes only the forward strand into account and the reverse strand has to be created manually or by a variety of software tools.
2. Primers should be approx 30 bp in length and the optimum size of an amplification product is approx 200 bp, although we also successfully analyzed DNA methylation in PCR products larger than 300 bp. Capture efficiency of the biotinylated amplification product also decreases with size.
3. Primers are to be designed to a region containing four or more cytosines that have been converted during bisulfite treatment to ensure that they are only complementary to completely converted DNA as the chemical treatment is rarely quantitative.

4. Primers should preferentially contain no CpG positions. If it cannot be avoided, the maximum number of CpG positions covered should be restricted to one, and this CpG position should not be included in the last five bases from the 3' terminus to avoid preferential amplification.
5. To ensure specificity, palindromes within primers and complementary sequences between primers, as well as degenerated bases and inosine should be avoided.

3.2.2. Design of Pyrosequencing Primers

Primers for Pyrosequencing can be designed using the web-based design software provided by Biotage. Note that this software was created for the purpose of single-nucleotide polymorphism (SNP) genotyping and some of the stringency criteria for design may be too strict for bisulfite-treated DNA.

1. Pyrosequencing primers can be designed on a simplified sequence in which all except the YpG position closest to the primer are replaced by either CpG or TpG.
2. A CpG position is chosen as the first position to be analyzed where a sequencing primer can be positioned without annealing to a second potential polymorphic position (SNP or CpG) in at least the last six bases from its 3' terminus.
3. Primers successfully designed by the software should be checked manually for primer dimers, possible hairpin structures, as well as further annealing possibilities. As few as four consecutive nucleotides complementary to a sequence in the amplification product might add to a background signal confounding the precise quantification.
4. The direction of the Pyrosequencing primer defines which of the amplification primers needs to be biotinylated. This primer should be checked carefully not to form any hairpin structure (*see* **Note 3**).
5. If no primers are found, an artificial polymorphism can be introduced at any base in the direction of the intended sequencing as the software positions a primer only up to around five bases away from the polymorphic target. This confinement is not necessary as with the PyroGold SQA kit a read-length of up to 80 bases can be obtained.
6. Successfully designed assays should be directly programmed into the Pyrosequencing software; depending on the used computer and the number of CpGs to be analyzed in a target sequence stretch, calculation times for the dispensation order may take up to several hours.

3.3. PCR Amplification

A specific, good-quality PCR product, clearly visible on an agarose gel after depositing 5 μL reaction mixture, needs to be obtained in order to ensure a successful Pyrosequencing reaction. PCR amplification should be carried out in a final volume of 20–25 μL, ideally in a 96-well plate using standard procedures, with one of the two PCR primers biotinylated (*see* **Notes 4** and **5**).

1. Amplification primers: 5 pmol of each primer are generally used for a 25-μL PCR (e.g., 0.2 μ*M*). An excess of primers should be avoided as the excess

biotinylated primer may diminish capture of the amplification product on the streptavidin-coated beads and may give rise to background signals during the Pyrosequencing reaction. Forty to 50 cycles of amplification should be performed to ensure complete exhaustion of the free biotinylated primer and a good yield of PCR product.

2. Template concentration: an increase in random preferential amplification during PCR is observed with decreasing amounts of template DNA used in the PCR amplification. This reduces reproducibility of the quantification quite drastically; standard deviation may be as large as 30%. A minimum of 10 ng of template DNA is necessary to achieve high reproducibility for each CpG methylation analysis.
3. DNA polymerases: so far we have not identified any DNA polymerases that are incompatible with the Pyrosequencing protocol. We routinely use either platinum *Taq* DNA polymerase (Invitrogen) or Qiagen's HotStar *Taq* DNA polymerase (*see* **Note 6**).
4. Appropriate controls should always be included, especially several negative controls as also very small amounts of contaminating DNA will be amplified owing to the high number of amplification cycles. Conventional decontamination methods such as dUTP incorporation are of course not applicable to bisulfite-treated DNA as nonmethylated cytosines are converted to dUTPs during bisulfite treatment (*see* **Note 7**).

3.4. Purification and Preparation

The amplified PCR products need to be rendered single stranded to enable annealing of the sequencing primer as the Pyrosequencing reaction takes place at 28°C because of thermal instability of the enzyme mix. Furthermore dNTPs have to be removed from the PCR reaction mixture to allow for controlled addition of single nucleotides. This preparation step is best carried out with a vacuum preparation tool (Biotage), which captures the beads and holds them during the different purification steps, whereas the solution easily passes through the filters ***(19)***. Template preparation can be applied to a complete 96-well plate as well as to a single well of a 96-well plate (*see* **Note 8**). One or several different Pyrosequencing primers can be used on the same plate (*see* **Note 9**). Purification is conducted directly on the beads fixed onto a filter after aspiration of the binding mix. The beads are then released in the final annealing buffer for the sequencing reaction.

1. Add 51 μL binding buffer and 4 μL streptavidin-coated sepharose beads to the 25-μL PCR product in each well of the PCR plate and incubate for 10 min at room temperature under constant mixing (1400 rpm). It is crucial not to allow the beads to pellet and/or precipitate before purification.
2. During this incubation step the Pyrosequencing plate can be prepared. 15 pmol Pyrosequencing primer in 40 μL of annealing buffer are dispensed into each well of the sequencing plate.

3. Fill troughs with 180 mL 70% ethanol, washing buffer, and water, respectively and the trough for denaturing with 120 mL of the denaturing solution. This different level assures that the NaOH is completely washed off as it otherwise might inhibit the ensuing reactions.
4. Create a vacuum in the aspiration device (450 mmHg) and aspirate the binding mix. This step is of utmost importance because if part of the beads stay in the wells an insufficient PCR product may be retrieved for a successful sequencing reaction.
5. Immerse the tips of the filters in three successive baths for 5 s each: ethanol 70%, denaturing solution, and washing solution. Turn over the tool and release the vacuum.
6. Immerse the tip of the filters in the annealing mix of the sequencing plate and shake gently to release the beads into the wells.
7. Incubate the sequencing plate for 2 min at 80°C on the thermoplate placed on a heating device. Sealing of the plate is not necessary. Allow the plate to cool at room temperature for annealing the primers to the PCR product.

3.5. Pyrosequencing Reaction

During cooling of the sequencing plate, program the sequencing run on the Pyrosequencing instrument with the assays entered beforehand (*see* **Subheading 3.2.2.**). Dispense the reagents (nucleotides, substrate, and enzyme mix) into the appropriate wells of the cartridge. Deposit the sequencing plate and the reagents' cartridge in the Pyrosequencing instrument and run the system.

For sequencing runs of up to 30 nt the PyroGold SNP reagent kit can be used. The PyroGold SQA reagent kit enables longer read-length of up to 90 nt. **Figure 3** shows typical Pyrogram charts for three samples analyzing six CpGs in the first exon of the human *HLA-G* gene. After the end of the run, the cartridge should be washed and deionized water pressed through the needles to verify unhindered and straight flow of the liquid for the next run.

The degree of methylation is calculated as allele frequency

$$\text{methylation \%} = \left[\frac{\text{peak height methylated}}{\text{peak height methylated} + \text{peak height non-methylated}}\right] * 100$$

using the allele quantification functionality of the PSQ 96MA software and can be exported to be further treated with statistical or graphical software.

4. Notes

1. The quality of genomic DNA preparation is essential for successful bisulfite conversion. The presence of residual amounts of proteins in the DNA sample is highly detrimental for complete bisulfite conversion. Treatment of the DNA with proteinase K during extraction is therefore recommended.

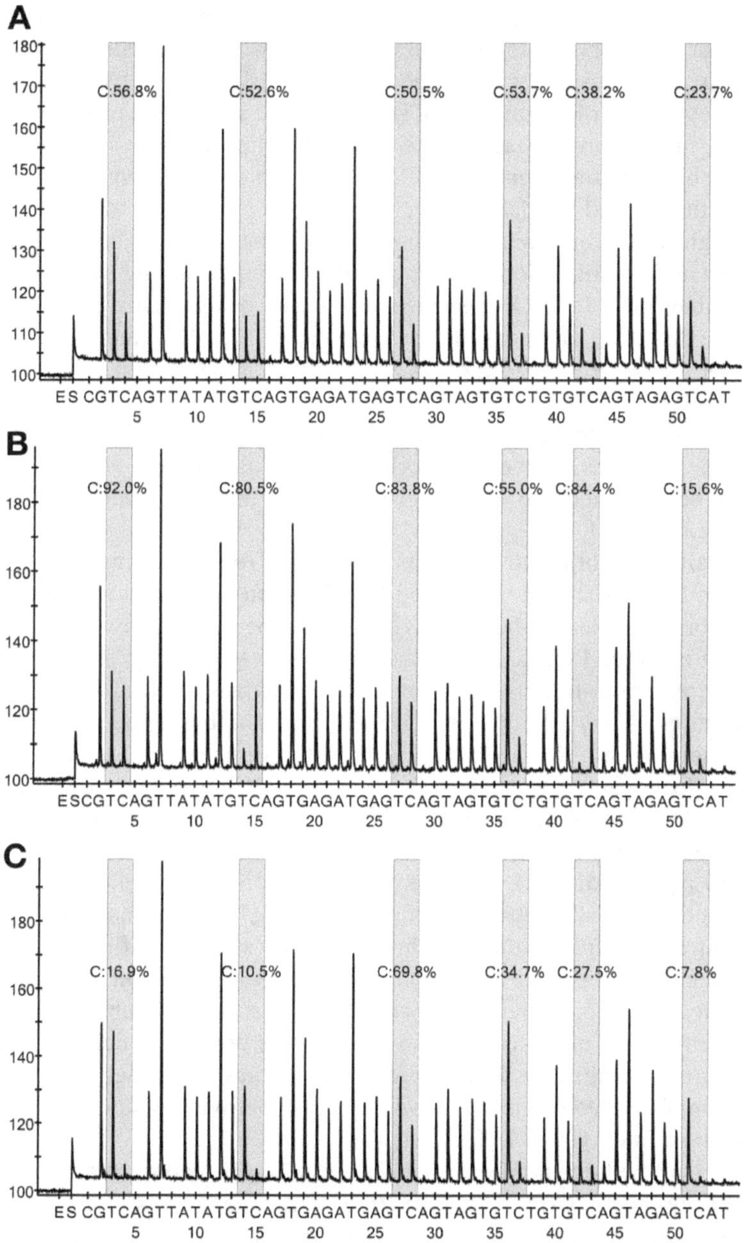

Fig. 3. Three Pyrogram® charts analyzing six CpGs in the first exon of the human *HLA-G* gene. (**A**) An average methylation degree of approx 50%, (**B**) approx 90%, and (**C**) approx 20%, although consecutive CpGs display a high variability of the methylation degree. Methods analyzing only single CpGs might have missed this quantitative difference.

2. The described protocol for bisulfite conversion is applied in routine use in our laboratory, but several modifications have been reported. Similar rates of conversion are obtained with incubation for only 4 h or by incubation in a thermocycler. For low amounts of starting DNA, a modified protocol embedding the DNA into agarose beads can be used ***(21)***. There are also several commercial kits for bisulfite treatment available on the market that might help implementation of the method in a laboratory.
3. Although the Pyrosequencing primer defines the strand to be amplified with a biotinylated amplification primer, we recommend designing the PCR amplification primer prior to the Pyrosequencing primer design. This procedure will ensure the detection of a possible secondary structure of the template, mispriming sites for the sequencing primer, as well as potential extension of the 3'-end of the template.
4. Biotinylated primers are more sensitive to storage conditions than unmodified primers. Dilutions, as well as stock tubes, should be kept at –20°C. Dilutions should be aliquoted and not subjected to more than five cycles of freezing and thawing.
5. The quality of biotinylated primers may differ from one commercial supplier to another. Primers should be ordered HPLC purified, as remaining free biotin occupies binding sites on the streptavidin-coated beads and may thereby reduce the yield of captured PCR product.
6. Commercial high-fidelity PCR kits do often contain archaeal DNA polymerases or mixtures of *Taq* DNA polymerase with archaeal DNA polymerases. Archaeal are efficiently blocked by dUTP leading to polymerase stalling. To minimize these effects some kits do contain dUTPase. PCR amplifications on bisulfite-treated DNA can therefore not be performed using any kit containing polymerases of archaeal origin such as *Pfu* or *Vent*.
7. A high number of CpG positions in the amplified fragment results in a large sequence difference between completely methylated and non-methylated template after bisulfite treatment. This difference strongly influences the melting and annealing properties during PCR amplification and might lead to preferential amplification of one allele (PCR bias); as it is sometimes observed in quantitative methylation analysis ***(22)***. As PCR bias is both sequence- and strand-specific, it is almost impossible to predict. PCR bias can either be detected using PCR-amplified DNA with a known degree of methylation or mixtures of methylated and unmethylated DNA.
8. Because of the large number of cycles, we strongly recommend a spatial separation of pre- and post-PCR manipulations to reduce the risk of contaminations.
9. Multiplexing of several primers in the same well is difficult, as mispriming of the Pyrosequencing may occur and quantification may be confounded as it may not be unambiguously assigned to one Pyrosequencing primer only. This lack of multiplexing abilities is compensated by the possibility to analyze up to 10 CpG positions with a single Pyrosequencing primer.

References

1. Bird, A. (2002) DNA methylation patterns and epigenetic memory. *Genes Dev.* **16,** 6–21.
2. Robertson, K. D. (2005) DNA methylation and human disease. *Nat. Rev. Genet.* **6,** 597–610.
3. Reik, W. and Walter, J. (2001) Genomic imprinting: parental influence on the genome. *Nat. Rev. Genet.* **2,** 21–32.
4. Maier, S. and Olek, A. (2002) Diabetes: a candidate disease for efficient DNA methylation profiling. *J. Nutr.* **132,** 2440S–2443S.
5. Laird, P. W. (2005) Cancer epigenetics. *Hum. Mol. Genet.* **14,** R65–R76.
6. Laird, P. W. (2003) The power and the promise of DNA methylation markers. *Nat. Rev. Cancer* **3,** 253–266.
7. Fraga, M. F. and Esteller, M. (2002) DNA methylation: a profile of methods and applications. *Biotechniques* **33,** 632–649.
8. Cottrell, S. E. and Laird, P. W. (2003) Sensitive detection of DNA methylation. *Ann. N. Y. Acad. Sci.* **983,** 120–130.
9. Ronaghi, M., Uhlen, M., and Nyren, P. (1998) A sequencing method based on real-time pyrophosphate. *Science* **281,** 363–365.
10. Uhlmann, K., Brinckmann, A., Toliat, M. R., Ritter, H., and Nürnberg, P. (2002) Evaluation of a potential epigenetic biomarker by quantitative methyl-single nucleotide polymorphism analysis. *Electrophoresis* **23,** 4072–4079.
11. Tost, J., Dunker, J., and Gut, I. G. (2003) Analysis and quantification of multiple methylation variable positions in CpG islands by Pyrosequencing. *Biotechniques* **35,** 152–156.
12. Colella, S., Shen, L., Baggerly, K. A., Issa, J. P., and Krahe, R. (2003) Sensitive and quantitative universal Pyrosequencing methylation analysis of CpG sites. *Biotechniques* **35,** 146–150.
13. Dupont, J. M., Tost, J., Jammes, H., and Gut, I. G. (2004) De novo quantitative bisulfite sequencing using the Pyrosequencing technology. *Anal. Biochem.* **333,** 119–127.
14. Yang, A. S., Estecio, M. R., Doshi, K., Kondo, Y., Tajara, E. H., and Issa, J. P. (2004) A simple method for estimating global DNA methylation using bisulfite PCR of repetitive DNA elements. *Nucleic Acids Res.* **32,** e38.
15. Jeong, K. S. and Lee, S. (2005) Estimating the total mouse DNA methylation according to the B1 repetitive elements. *Biochem. Biophys. Res. Commun.* **335,** 1211–1216.
16. Issa, J. P., Gharibyan, V., Cortes, J., et al. (2005) Phase II study of low-dose decitabine in patients with chronic myelogenous leukemia resistant to imatinib mesylate. *J. Clin. Oncol.* **23,** 3948–3956.
17. Shiao, Y. H., Crawford, E. B., Anderson, L. M., Patel, P., and Ko, K. (2005) Allele-specific germ cell epimutation in the spacer promoter of the 45S ribosomal RNA gene after Cr(III) exposure. *Toxicol. Appl. Pharmacol.* **205,** 290–296.
18. Li, L. C. and Dahiya, R. (2002) MethPrimer: designing primers for methylation PCRs. *Bioinformatics* **18,** 1427–1431.

19. Dunker, J., Larsson, U., Petersson, D., et al. (2003) Parallel DNA template preparation using a vacuum filtration sample transfer device. *Biotechniques* **34,** 862–868.
20. Frommer, M., McDonald, L. E., Millar, D. S., et al. (1992) A genomic sequencing protocol that yields a positive display of 5-methylcytosine residues in individual DNA strands. *Proc. Natl. Acad. Sci. USA* **89,** 1827–1831.
21. Olek, A., Oswald, J., and Walter, J. (1996) A modified and improved method for bisulphite based cytosine methylation analysis. *Nucleic Acids Res.* **24,** 5064–5066.
22. Warnecke, P. M., Stirzaker, C., Melki, J. R., Millar, D. S., Paul, C. L., and Clark, S. J. (1997) Detection and measurement of PCR bias in quantitative methylation analysis of bisulphite-treated DNA. *Nucleic Acids Res.* **25,** 4422–4426.

9

Detecting Mutations That Confer Oxazolidinone Resistance in Gram-Positive Bacteria

Neil Woodford, Sarah E. North, and Matthew J. Ellington

Summary

Resistance to oxazolidinone antibiotics, including linezolid, in Gram-positive bacteria is mediated by single-nucleotide polymorphisms (SNPs) in the 23*S* ribosomal RNA. A G2576U change (encoded by a G2576T mutation in the rRNA genes) is found in most resistant clinical isolates of enterococci and staphylococci; a variety of changes have been found in resistant mutants selected in vitro. Pyrosequencing® can be used to detect SNPs known to confer oxazolidinone resistance, including the G2576T change. Most bacteria have more than one rRNA gene copy and Pyrosequencing can also be used for allele quantification, i.e., to estimate the proportions of mutant vs wild-type alleles. The number of mutated rRNA gene copies correlates roughly with the level of oxazolidinone resistance displayed by resistant isolates. This chapter summarizes the Pyrosequencing assays that have been developed in our laboratory for analyzing oxazolidinone-resistant enterococci and staphylococci.

Key Words: Antibiotic resistance; oxazolidinone; linezolid; 23*S* rRNA; single-nucleotide polymorphism; SNP detection; allele quantification.

1. Introduction

The oxazolidinones are the first new class of antibacterials developed and launched into clinical use for over 30 yr, with linezolid the first licensed member of the class. Oxazolidinones have a novel mechanism of action, preventing protein synthesis by inhibiting formation of the 70*S* ribosomal initiation complex, and prevalent mechanisms of resistance to other protein synthesis-inhibiting antibiotics do not confer cross-resistance to them ***(1–3)***. The oxazolidinones are synthetic agents and there is no known pre-existing reservoir of resistance genes, but experience has warned microbiologists and clinicians against complacency; bacteria always develop resistance to any new

From: *Methods in Molecular Biology, vol. 373: Pyrosequencing® Protocols*
Edited by: S. Marsh © Humana Press Inc., Totowa, NJ

antibiotic. Indeed, a few linezolid-resistant enterococci were selected during prelicensing clinical trials ***(4)***, and resistance has emerged, rarely, postlicensing in enterococci ***(5–8)*** and *Staphylococcus aureus* ***(9–11)***, mostly in patients who were underdosed, had difficult-to-reach infections, or who required long treatment. Resistance is mediated by single-nucleotide polymorphisms (SNPs) in the genes that encode 23*S* ribosomal RNA; several have been reported in laboratory-generated mutants ***(12)***, but a G2576T change is present in most (though not all) resistant clinical isolates of enterococci and staphylococci ***(5–11)***.

Oxazolidinone resistance is difficult to select in vitro, and quantification of mutation frequencies is often not possible. However, linezolid-resistant mutants can be raised by serial passage of bacteria in the presence of increasing concentrations of the drug. The low mutation frequency arises, in part, because the target gene, 23*S* rDNA, is present in multiple copies in most bacterial species. In consequence, a single mutational event in one gene copy generally is insufficient to confer phenotypical linezolid resistance. It seems likely that subsequent intrachromosomal recombination (also known as gene conversion ***[13]***) is needed to distribute the altered gene copies (or relevant fragments of them) to multiple rRNA genes. In support of this, there is correlation between the number of rRNA gene copies carrying known linezolid-resistance SNPs and the level of resistance (measured as the minimum inhibitory concentration [MIC]) of linezolid ***(14–16)***. Also consistent with this two-step theory of resistance development, linezolid-resistant mutants were selected at lower frequencies from a recombination-deficient *Enterococcus faecalis* strain than from a partially isogenic wild-type strain ***(17)***.

Currently, linezolid resistance is not common, though it emerges as a potential problem in patients who receive prolonged courses of treatment. Only time can determine whether linezolid-resistant Gram-positive cocci will become a major public health concern. If rare chromosomal mutation remains the only resistance mechanism, we need to be more concerned about the potential for spread of the linezolid-resistant strains, rather than the *de novo* selection of resistance genes; a hospital cluster of infections caused by linezolid-resistant enterococci affecting eight patients on an oncology ward has been reported ***(18)***.

Rapid detection of linezolid-resistant isolates is essential for the correct management of individual patients, and for controlling spread of the resistant bacteria. This can be achieved by phenotypical susceptibility testing in routine bacteriology laboratories, but should be confirmed by a reference laboratory, whenever possible. Ideally, these reference centers should also investigate the resistance mechanism in all referred isolates so that any novel mechanisms are recognized rapidly. Any method suitable for SNP detection can be adapted to study linezolid resistance, and those published include traditional sequencing,

PCR-RFLP analysis, and real-time PCR *(10,16,19,20)*. Pyrosequencing has been used in our laboratory, alongside some of these other methods, to detect the G2576T SNP in linezolid-resistant clinical isolates of enterococci *(16)* and, more recently, to detect G2576T and other SNPs in linezolid-resistant clinical isolates and laboratory-generated mutants of *S. aureus*. This chapter summarizes these assays; a fuller description of the principles of Pyrosequencing can be found elsewhere in this volume *(21)*. The assays described could be adapted, with comparatively little effort, to allow detection of SNPs causing oxazolidinone resistance in coagulase-negative staphylococci and streptococci.

For a laboratory seeking to investigate the genetic basis of resistance in a single linezolid-resistant isolate, Pyrosequencing is not a cost-effective approach. PCR-RFLP assays are often more appropriate for small numbers of isolates, and can be used to detect the SNPs, G2576T (*Nhe*I) *(19)*, G2505A (*Eco*RV) *(17)*, or T2504C (*Hin*1I) (S. E. North, unpublished data). However, the SNPs G2445T, G2447T, T2500A, and A2503G do not result in the creation or loss of restriction sites and cannot be detected by PCR-RFLP analysis; sequencing is required for their detection. Pyrosequencing is an ideal method for larger laboratories that seek to process larger numbers of oxazolidinone-resistant isolates, and where quantification of the number of mutant vs wild-type 23*S* rRNA genes may yield useful data.

2. Materials

1. Pyrosequencing assay design software (Biotage AB, Uppsala, Sweden; *see* **Note 1**).
2. Standard microbiological media/equipment for growing the bacteria of interest.
3. Thermal cycler and standard PCR reagents for amplifying the required target.
4. Target-specific primer pair; one primer in each pair should be manufactured with a 5'-biotin label (*see* **Subheading 3.1.**; **Notes 2** and **3**) (**Table 1**). Primers should be dissolved in PCR-quality water to a stock concentration of 1 μg/μL.
5. Molecular biology-grade agarose, 0.5X Tris-borate EDTA buffer pH 8.0, DNA sub-cell and power supply for electrophoresis, ethidium bromide and UV transilluminator *(22)*.
6. Biotinylated strand capture: binding buffer (Biotage AB); streptavidin-sepharose HP (e.g., Amersham Biosciences AB, Uppsala, Sweden, cat. no. 17-5113-01).
7. Captured strand cleaning: vacuum prep workstation (Biotage AB) and "clean-up" buffers: denaturation solution; washing buffer (Biotage AB); also 70% ethanol and high-purity water (**Fig. 1A**).
8. Sequencing primer (*see* **Note 4**). Primers should be dissolved in PCR-quality water to a stock concentration of 1 μg/μL.
9. Primer annealing buffer (Biotage AB).
10. Pyrosequencing instrument (e.g., PSQ 96MA; Biotage AB).
11. Pyro-Gold reagents (e.g., 5X 96 PSQ 96MA kit; Biotage AB) and loading cartridge (**Fig. 1B**; *see* **Note 5**).

Table 1
Primers Suitable for Detecting SNPs Responsible for Oxazolidinone Resistance in Enterococci and Staphylococci by Pyrosequencing

Mutation in 23*S* rRNA gene[a]	Assay suitable for	Amplification primers[b]	Pyrosequencing primer
G2576T (assay I)	Clinical isolates and laboratory mutants, enterococci, and staphylococci	F 5'-Biotin-GCA TCC TGG GGC TGT AGT C R 5'-GGA CCG AAC TGT CTC ACG AC ***(16)***	5'-CGT TCT GAA CCC AGC ***(16)***
G2576T (assay II)	Clinical isolates and laboratory mutants of staphylococci[c]	F 5'-AAG GGC CAT CGC TCA ACG GAT R 5'-<u>CCGAATAGGAACGTTGAGCCGT</u> GAG GGG GGC TTC ATG CTT Universal primer: 5'-Biotin-CCG AAT AGG AAC GTT GAG CCG T	5'-TTCGCCCATTAAAGCGGTAC[c]
T2500A A2503G T2504C G2505A	Clinical isolate of *Staphylococcus aureus* (T2500A); laboratory mutants of *S. aureus* (A2503G, T2504C); laboratory mutants of *S. aureus* and enterococci (G2505A)	As for G2576T (assay II)	5'-ACG GGG AGG TTT GGC AC
G2445T G2447T	Laboratory mutants of *S. aureus*[d]	As for G2576T (assay II)	5'-AACGGATAAAAGCTACCCCG[d]

[a]Positions are numbered with reference to the 23*S* rRNA sequence of *Escherichia coli*, e.g., GenBank AF053964.

[b]G2576T (assay I) illustrates Pyrosequencing with a target-specific biotinylated primer ***(16)***. The other assays illustrate an alternative approach in which one target-specific amplification primer includes a universal handle sequence (underlined). This strategy allows a universal biotinylated primer to be used, thereby reducing costs in laboratories running several Pyrosequencing assays (*see* **Note 2**).

[c]The Pyrosequencing primer has been shown to detect the G2576T SNP in *S. aureus*, but has a single mismatch with the 23*S* rDNA sequence of *Enterococcus faecalis* and *E. faecium*. The corrected primer sequence for enterococci (not evaluated by the authors) would be: 5'-TTCGCCCATTAAAGCGGCAC.

[d]The Pyrosequencing primer has been shown to detect the G2445T and G2447T SNPs in *S. aureus*, but has a single mismatch with the 23*S* rDNA sequence of *E. faecalis* and *E. faecium*. The corrected primer sequence for enterococci (not evaluated by the authors) would be: 5'-AACGGATAAAAGCTACCCTG.

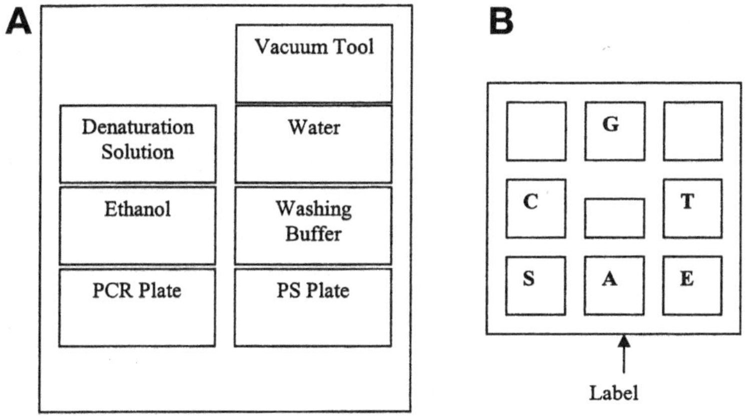

Fig. 1. (**A**) The vacuum prep workstation for clean up of the amplified PCR. The PCR plate contains amplified PCR products that already have been mixed with streptavidin beads. After processing, the single-stranded DNA templates will be located in the Pyrosequencing plate, prior to transfer to the Pyrosequencing® instrument. (**B**) The reservoirs of the Pyrosequencing cartridge are each filled with the appropriate proprietary reagent needed for the assay: S, substrates; E, enzyme mix; A, α-thio-dATP; C, dCTP; G, dGTP; and T, dTTP.

3. Methods

3.1. Amplification of Pyrosequencing Template DNA

1. Proprietary software developed specifically for designing amplification and Pyrosequencing primers can be purchased from Biotage AB (*see* **Notes 1–4**).
2. Amplify the target region of interest using suitable PCR primers (**Table 1**).
3. Make up primer mastermix with 5 μL forward primer, 5 μL reverse primer, 10 μL universal biotinylated primer, and 80 μL PCR-quality water.
4. Conditions suitable for the assays described here are initial denaturation at 94°C for 5 min, 30 cycles of 94°C for 25 s, 52°C for 40 s, and 72°C for 50 s; and a final elongation at 72°C for 6 min.
5. Confirm the specificity of the PCR by electrophorezing 4 μL of the PCR product on a 2% agarose gel. Stain with ethidium bromide (final concentration, 1 μg/mL) and visualize under UV light.

3.2. Post-PCR Processing of the Product (Fig. 2)

1. Prepare a suitable number of wells in a PCR plate (one well per PCR product for Pyrosequencing). Each well should contain 30 μL PCR product, 37 μL binding buffer, 3 μL streptavadin-sepharose bead mix, and 10 μL PCR-quality water.
2. Cover the plate with sealing tape or strip caps, and agitate at 1400 rpm/3 mm mixing stroke (Eppendorf thermomixer comfort MTP) for 10 min at room temperature.

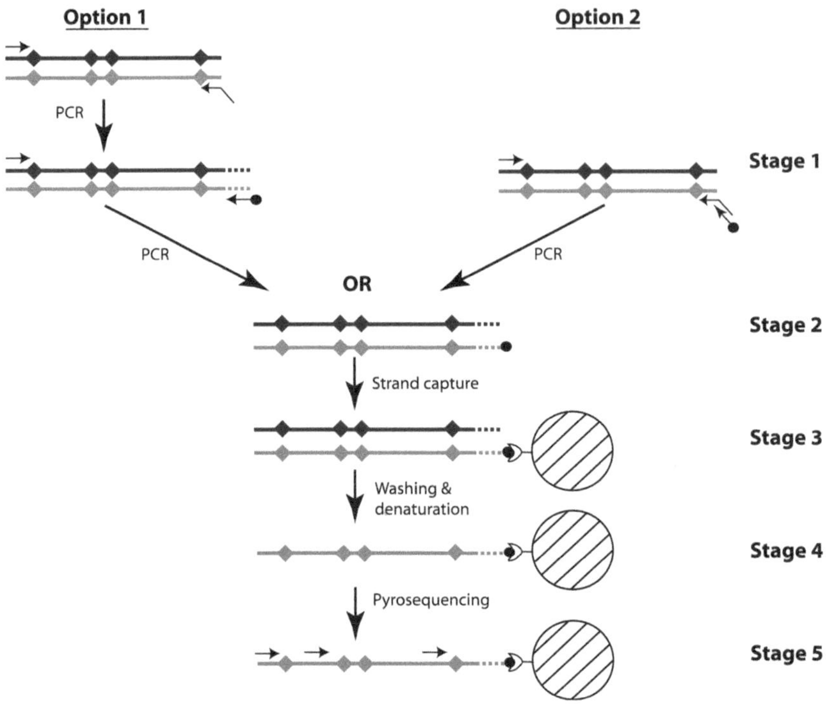

Fig. 2. Schematic to illustrate the Pyrosequencing® process with a universal biotinylated PCR primer; adapted from **ref. *23***. Stage 1: when optimizing assays (option 1), the target sequence is initially amplified with a pair of target-specific primers. In the example, the reverse primer has a universal handle sequence at the 5'-end. The primary PCR product, which incorporates the handle sequence, is then used as template and reamplified in a second round of PCR with the target-specific forward primer and a universal, biotinylated reverse primer. For established, optimized assays (option 2) all three primers can be combined in a single round of PCR, which reduces assay time. Stage 2: both of these options generate the required PCR product with a biotin label incorporated into the reverse strand. Stage 3: the biotinylated PCR product is captured on streptavidin-sepharose, denatured, and washed. Stage 4: the resulting single-stranded DNA is ready to use as template in the Pyrosequencing reaction. Stage 5: Pyrosequencing is undertaken with a target-specific primer using proprietary reagents and equipment.

3. Prepare a Pyrosequencing (PS) plate, again with one well per PCR product for Pyrosequencing. Each well should contain 44 μL annealing buffer and 1 μL of the appropriate Pyrosequencing primer.
4. Strand separation is achieved using a vacuum prep tool and workstation. Initially, fill the troughs in the workstation with 180 mL of the solutions outlined in **Fig. 1A**, and refill as required.

5. Connect the vacuum tool to a vacuum pump and switch it on.
6. Wash the pins of the vacuum tool with high-purity water for 30 s, place the vacuum tool into the PCR plate, and ensure the liquid has been aspirated evenly from all wells. The sepharose beads with bound PCR product will stick to the filters of the tool's pins. The surface of the remaining solution should be lower than, or even with, the pin tips when they are resting on the bottom of the well.
7. Wash the pins with 70% ethanol for 5 s.
8. Wash the pins with denaturation solution for 5 s.
9. Wash the pins with washing buffer for 5 s.
10. Release the vacuum and lower the pins into the PS plate (**Fig. 1A**), the wells of which contain annealing buffer and Pyrosequencing primer. Release the beads with single-stranded PCR product into the annealing buffer.
11. Wash the vacuum tool with high-purity water for 10 s to release any remaining beads.
12. If preparing more than one plate, place the pins into the new PCR plate and proceed as before.
13. When all plates have been processed, rinse the vacuum tool for 20–30 s with high-purity water. Store the vacuum tool dry.

3.3. Pyrosequencing

1. Heat the PS plate for 2 min at 80°C and cool to room temperature to allow the Pyroseqencing primer to anneal to the single-stranded template DNA. While it is cooling, progress to **steps 2–8**.
2. Set up a run on the computer and select the appropriate processing mode, either SNP (for SNP analysis) or SQA (for sequencing). For SNP mode, the position of the SNP(s) to be studied is programmed into the Pyrosequencing instrument (*see* **Note 6**).
3. Reconstitute the enzyme and substrate mixtures (each with 620 μL of high-purity water; *see* **Note 5**).
4. Calculate the volume of enzyme required. Use 50 μL + (5.5 μL × no. of wells) for a partially filled plate. Use the total volume (620 μL) for a full plate (96 samples).
5. The volume of substrates used is the same as the volume of enzyme.
6. Calculate the volume of nucleotides required. There is a maximum of seven dispensations of each nucleotide per well. Hence, use 50 μL + (0.11 μL × 7 × no. of wells). If a full plate (96 wells) is being processed, use 100 μL of nucleotides.
7. Dilute the calculated volume of nucleotides with an equal volume of TE buffer. Use 100 μL TE for a full plate (*see* **Note 7**).
8. Place a reagent cartridge with the label facing you. Fill with appropriate amounts of enzyme, substrate, dCTP, dGTP, dTTP, and α-thio-dATP (**Fig. 1B**). Allow the solutions and cartridge to reach room temperature before starting the run.
9. Transfer the PS plate and cartridge to the Pyrosequencing machine, and start the run.
10. The machine will dispense reagents sequentially in accordance with the program selected.

11. The user will be alerted by the computer at the end of the program and data can then be analyzed.
12. At the end of the run, wash the reagent cartridge with high-purity water, forcing water through it (*see* **Note 8**).

3.4. Analysis of Data

1. Ideally, all Pyrosequencing assays should be run with control isolates that have been shown to contain the SNPs under investigation by other methods, such as traditional sequencing or PCR-RFLP analysis.
2. Pyrosequencing generates data in the form of a Pyrogram® (**Fig. 3**), which shows relative peak heights (equivalent to bursts of light) following the addition of each nucleotide; the amount of light emitted and the peak height is proportional to the amount of nucleotide added.
3. The sequence of a test DNA fragment may be read from the Pyrogram, peaks of "double height" or larger indicate addition of the particular nucleotide in two or more adjacent positions (*see* **Note 9**). The software automatically calls sequence, which may be checked against the Pyrogram.
4. If the Pyrosequencing software was programmed for a SNP run, all nucleotides added in the position of the SNP are highlighted yellow by the software during SNP analysis for ease of recognition; no shading is used if the machine was used in SQA (sequencing) mode.
5. If you select the allele quantification option in the software, it will return figures for the proportion of each nucleotide added at the specified SNP position. This can be used together with a (known) number of alleles present in the isolate or species under investigation to estimate the numbers of wild-type vs mutated alleles (*see* **Note 10**).
6. Biotage AB markets software that can import raw sequence data directly from the Pyrosequencing instrument and can compare it with local user-defined database(s) of reference sequences (in FASTA format). This "Identifire" software is being marketed for identification, but it has clear potential for any application.

Fig. 3. *(opposite page)* The "Pyrogram®. Detection by Pyrosequencing® of the G2576T mutation in 23*S* rRNA genes of linezolid-resistant and -susceptible enterococci with assay I (**Table 1**), which yields sequence for the reverse strand: **(A)** a homozygous susceptible isolate with wild-type sequence (G2576) in all gene copies (a complementary C is incorporated); **(B)** a homozygous resistant isolate with a mutation (T2576) in all gene copies (a complementary A is incorporated); **(C)** a heterozygous resistant isolate with both mutant and wild-type copies of the gene (complementary C and A are both incorporated). The shaded region shows the SNP at position 2576, which was programmed into the Pyrosequencing software. Specified dispensation order for nucleotides is shown along the "*x*-axis" of the figure. The initial additions of enzymes and substrates to the reaction mix are indicated by E and S, respectively. (Reproduced from **ref. *16*** with permission of the American Society for Microbiology.)

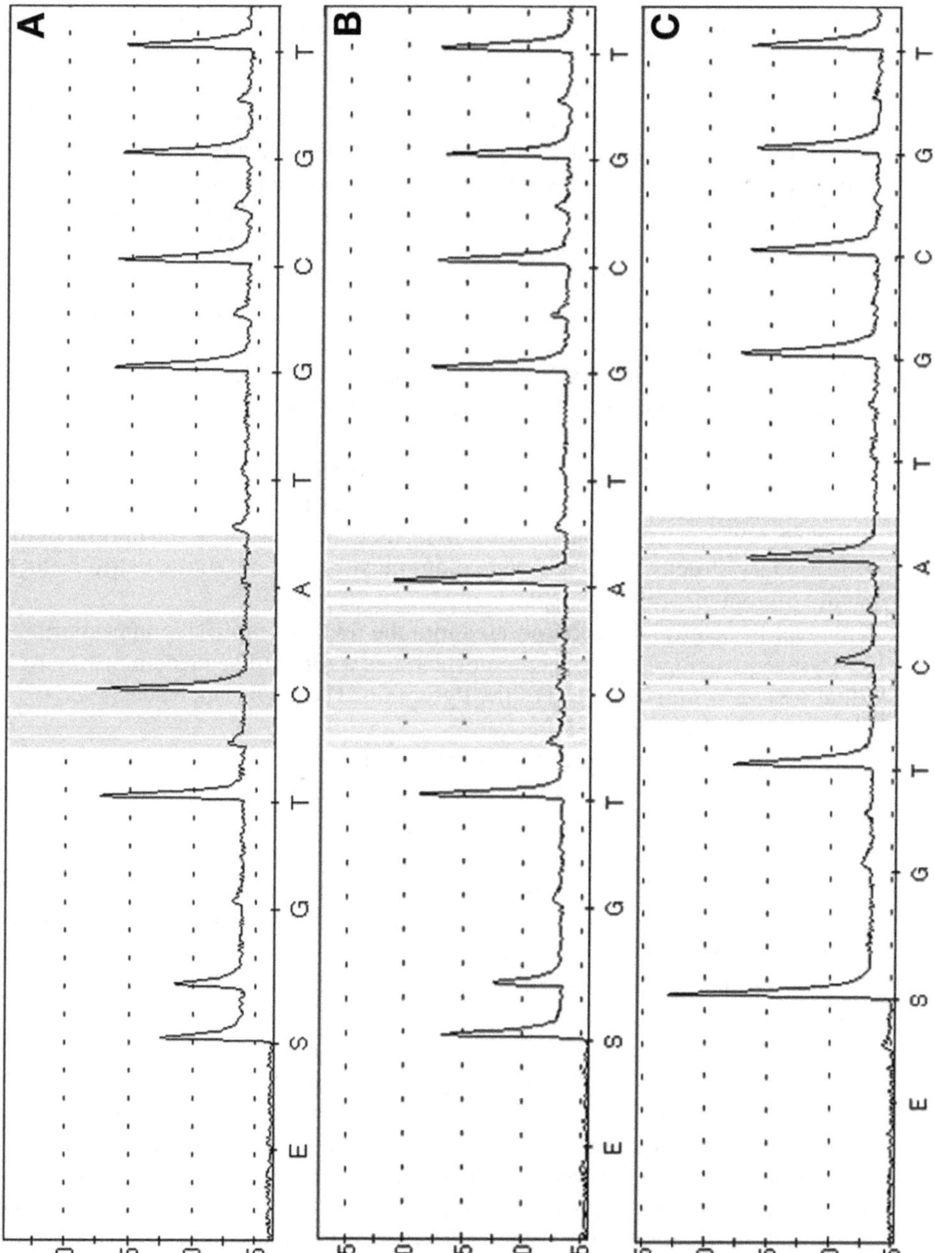
A
B
C
E S G T C A T G C G T
E S G T C A T G C G T
E S G T C A T G C G T
105 110 115 120
105 110 115 120 125
105 110 115 120 125

4. Notes

1. Details of the full range of Pyrosequencing equipment, consumables, design and analysis software, and technical support can be found at http://www.biotagebio.com.
2. The cost of Pyrosequencing can be reduced if a universal handle sequence is included at the 5'-end of either the target-specific forward or reverse primer *(22)* (**Table 1**). A third biotinylated oligonucleotide with a sequence identical to this universal handle can then be used to generate the required biotinylated PCR product. The three primers may be included in a single PCR reaction, or the biotinylated primer may be used in a second round of PCR (the latter option obviously increases the processing time) (**Fig. 1**). This adaptation avoids the need for biotinylated assay-specific primers, with only one biotinylated primer being necessary. A sequence suitable for use as the universal handle would be the 22-mer, 5'-CCG AAT AGG AAC GTT GAG CCG T-3' (M. Ronaghi, personal communication); addition of a biotin tag at the 5'-end will create the universal biotinylated oligonucleotide (**Table 1**).
3. The melting temperature (T_m) for the target-specific portion of the primer must be compatible not only with its target-specific partner but also with the handle sequence. Other usual considerations for primer design, such as the minimization of secondary structure and self-annealing, must be applied to the entire primer-handle oligonucleotide.
4. If the handle sequence is incorporated into the reverse amplification primer then the Pyrosequencing primer can be designed as a forward sequencing primer facilitating a direct forward read of the Pyrogram. The 3' OH can be designed to be adjacent to the first target SNP.
5. Store nucleotides and freeze-dried enzyme and substrate (light-sensitive) mixtures at 4–8°C. Do not exceed the expiry date of the kit. Freeze-dried reagents are reconstituted by adding 620 μL of high-purity water to the vials supplied. When dissolved, the reagents are stable for at least 5 d at 4–8°C. Alternatively, to minimize loss of activity the reconstituted enzyme and substrate mixtures can be aliquoted and stored at –20°C. Once thawed these reagents should *not* be refrozen. The nucleotides must *never* be frozen.
6. When programing a Pyrosequencing instrument for SNP mode the user is prompted to specify the nucleotide dispensation order. It is possible to use repeated cycles of, e.g., (G, A, T, C,) (G, A, T, C,) and so on, but this increases the run time. However, if the SNP is located in a region where all adjacent nucleotides are conserved, it is possible to specify the exact sequence of the flanking DNA. This avoids waste of nucleotides. The position of the SNP is programmed into the Pyrosequencing instrument, together with the nucleotides to be dispensed; if a SNP has complete degeneracy, then (G, A, T, C) will need to be dispensed; in the case of the G2576T SNP, A or C are not found at this position and do not need to be dispensed. This dispensation "streamlining" for SNP analysis speeds up Pyrosequencing markedly.

7. Diluted nucleotides are only stable for up to 24 h. Only prepare the required amount for one plate per batch at a time. The nucleotides must *never* be frozen.
8. It is recommended that reagent delivery cartridges are not used more than 20 times. This can be done simply by placing a fingertip over the reservoir end of the cartridge after it has been filled with water.
9. Care must be taken when interpreting Pyrogram charts that include multiples of particular nucleotides.
10. During analysis the Pyrosequencing software color codes the quality of returned data; blue for "passed," red for "failed" and orange for "check." The Pyrogram charts should be checked carefully for all "red" or "orange" returns as it may be possible to call the sequence visually.

References

1. Ford, C. W., Hamel, J. C., Stapert, D., et al. (1997) Oxazolidinones: new antibacterial agents. *Trends Microbiol.* **5,** 196–200.
2. Shinabarger, D. (1999) Mechanism of action of the oxazolidinone antibacterial agents. *Expert. Opin. Investig. Drugs* **8,** 1195–1202.
3. Ford, C. W., Zurenko, G. E., and Barbachyn, M. R. (2001) The discovery of linezolid, the first oxazolidinone antibacterial agent. *Curr. Drug Targets Infect. Disord.* **1,** 181–199.
4. Zurenko, G. E., Todd, W. M., Hafkin, B., et al. (1999) Development of linezolid-resistant *Enterococcus faecium* in two compassionate use program patients treated with linezolid. *Proc. abstr. 39th Intersci. Conf. Antimicrob. Agents Chemother.* American Society for Microbiology, p. 118.
5. Gonzales, R. D., Schreckenberger, P. C., Graham, M. B., Kelkar, S., DenBesten, K., and Quinn, J. P. (2001) Infections due to vancomycin-resistant *Enterococcus faecium* resistant to linezolid. *Lancet* **357,** 1179
6. Auckland, C., Teare, L., Cooke, F., et al. (2002) Linezolid-resistant enterococci: report of the first isolates in the United Kingdom. *J. Antimicrob. Chemother.* **50,** 743–746.
7. Johnson, A. P., Tysall, L., Stockdale, M. W., et al. (2002) Emerging linezolid resistant *Enterococcus faecalis* and *Enterococcus faecium* isolated from two Austrian patients in the same intensive care unit. *Eur. J. Clin. Microbiol. Infect. Dis.* **21,** 751–754.
8. Halle, E., Padberg, J., Rosseau, S., Klare, I., Werner, G., and Witte, W. (2004) Linezolid-resistant *Enterococcus faecium* and *Enterococcus faecalis* isolated from a septic patient: report of first isolates in Germany. *Infection* **32,** 182–183.
9. Tsiodras, S., Gold, H. S., Sakoulas, G., et al. (2001) Linezolid resistance in a clinical isolate of *Staphylococcus aureus. Lancet* **358,** 207–208.
10. Wilson, P., Andrews, J. A., Charlesworth, R., et al. (2003) Linezolid resistance in clinical isolates of *Staphylococcus aureus. J. Antimicrob. Chemother.* **51,** 186–188.
11. Meka, V. G., Pillai, S. K., Sakoulas, G., et al. (2004) Linezolid resistance in sequential *Staphylococcus aureus* isolates associated with a T2500A mutation

in the 23S rRNA gene and loss of a single copy of rRNA. *J. Infect. Dis.* **190,** 311–317.

12. Prystowsky, J., Siddiqui, F., Chosay, J., et al. (2001) Resistance to linezolid: characterization of mutations in rRNA and comparison of their occurrences in vancomycin-resistant enterococci. *Antimicrob. Agents Chemother.* **45,** 2154–2156.
13. Klein, H. L. (1995) Genetic control of intrachromosomal recombination. *Bioessays* **17,** 147–159.
14. Marshall, S. H., Donskey, C. J., Hutton-Thomas, R., Salata, R. A., and Rice, L. B. (2002) Gene dosage and linezolid resistance in *Enterococcus faecium* and *Enterococcus faecalis. Antimicrob. Agents Chemother.* **46,** 3334–3336.
15. Ruggero, K. A., Schroeder, L. K., Schreckenberger, P. C., Mankin, A. S., and Quinn, J. P. (2003) Nosocomial superinfections due to linezolid-resistant *Enterococcus faecalis*: evidence for a gene dosage effect on linezolid MICs. *Diagn. Microbiol. Infect. Dis.* **47,** 511–513.
16. Sinclair, A., Arnold, C., and Woodford, N. (2003) Rapid detection and estimation by Pyrosequencing of 23S rRNA genes with a single nucleotide polymorphism conferring linezolid resistance in enterococci. *Antimicrob. Agents Chemother.* **47,** 3620–3622.
17. Lobritz, M., Hutton-Thomas, R., Marshall, S., and Rice, L. B. (2003) Recombination proficiency influences frequency and locus of mutational resistance to linezolid in *Enterococcus faecalis. Antimicrob. Agents Chemother.* **47,** 3318–3320.
18. Engemann, J. J., Joyce, M. J., Harrell, L. J., et al. (2003) Outbreak of linezolid-resistant *Enterococcus faecium* bloodstream infections on an oncology ward. *Proc. abstr. 43rd Intersci. Conf. Antimicrob. Agents Chemother.* American Society for Microbiology, p. 359.
19. Woodford, N., Tysall, L., Auckland, C., et al. (2002) Detection of oxazolidinone-resistant *Enterococcus faecalis* and *Enterococcus faecium* strains by real-time PCR and PCR-restriction fragment polymorphism analysis. *J. Clin. Microbiol.* **40,** 4298–4300.
20. Werner, G., Strommenger, B., Klare, I., and Witte, W. (2004) Molecular detection of linezolid resistance in *Enterococcus faecium* and *Enterococcus faecalis* by use of 5' nuclease real-time PCR compared to a modified classical approach. *J. Clin. Microbiol.* **42,** 5327–5331.
21. Nyren, P. (2007) The history of Pyrosequencing. In: *Pyrosequencing Protocols*, (Marsh, S., ed.), Humana Press, Totowa, NJ, pp.1–14.
22. Johnson, A. P. and Woodford, N. (1998) Plasmid analysis. *Methods Mol. Med.* **15,** 51–62.
23. Fakhrai-Rad, H., Pourmand, N., and Ronaghi, M. (2002) Pyrosequencing: an accurate detection platform for single nucleotide polymorphisms. *Hum. Mutat.* **19,** 479–485.

10

Pyrosequencing®-Based Strategies for Improved Allele Typing of Human Leukocyte Antigen Loci

Steven Ringquist, Alexis Styche, William A. Rudert, and Massimo Trucco

Summary

Successful transplantation of tissue during solid organ and bone marrow transplantation relies on accurate determination of the human leukocyte antigen (*HLA*) phenotype of the potential donor(s) and recipient. Matching donor with recipient for a kidney transplant generally means finding a six-antigen match by looking at each of two alleles at *HLA-A*, *-B*, and *-DR* loci. For bone marrow transplantation the *HLA-C* and *-DQ* alleles are also considered. Molecular techniques, including sequencing, are capable of precisely defining *HLA* alleles. Because of the large number of possible allelic combinations there are numerous ambiguities associated with heterozygous genotypes even when sequence-based typing protocols are used. Sequencing-by-synthesis methodology employed by Pyrosequencing® represents an improvement when applied to *HLA* genotyping that allows resolution of many ambiguous allelic pairs. Out-of-phase sequencing of *HLA* alleles by Pyrosequencing can resolve *cis/trans* ambiguities that would otherwise require the sequencing of isolated cloned DNAs. Single-nucleotide polymorphism typing of *HLA* for the presence of specific variants is also beneficial for monitoring *HLA*-encoded genetic risk to autoimmune diseases, such as celiac disease, rheumatoid arthritis, and type 1 diabetes mellitus.

Key Words: Histocompatibility antigens; MHC; sequence-based typing; transplantation.

1. Introduction

The genes encoding the human leukocyte antigen (*HLA*) class I and class II proteins constitute the most highly polymorphic genetic system in humans. These loci represent stable polymorphisms that give rise to inherited alleles, i.e., alternative forms of the gene products. Allelic variations have been identified for most of the *HLA* loci, with *HLA*-B being the most polymorphic having more than 600 reported alleles and averaging 86 single-nucleotide polymor-

From: *Methods in Molecular Biology, vol. 373: Pyrosequencing® Protocols*
Edited by: S. Marsh © Humana Press Inc., Totowa, NJ

phisms (SNPs) per kilobase ***(1)***. The numbers of confirmed *HLA* alleles is consistently rising ***(2,3)***, there being as many as 1293 and 733 alleles reported for the class I and class II *HLA* loci, respectively.

HLA class I and class II loci are located in a more than 4-mb long DNA segment on the short arm of chromosome 6 at cytogenetic location 6p21.3 ***(1,4)***. The *HLA*-encoding region can be subdivided into three physical regions encoding the class I, class III, and class II molecules ***(5)***. The physiological function of the *HLA* proteins is to present antigenic peptides from the cytoplasm onto the cell surface. Knowledge of the identity of the class I and class II loci alleles are most important in the context of histocompatibility. Class I genes are encoded at the *A*, *B*, and *C* loci, whereas class II genes encode *HLA-DQB* and *HLA-DRB*. The class I genes and the class II gene *HLA-DQB1* are represented by one locus in each individual. In contrast, depending on the identity of the *HLA-DRB1* allele other class II *HLA-DRB* loci may also be present (**Fig. 1**). For example, the *HLA-DRB1* alleles *01 are in linkage disequilibrium with pseudogenes *HLA-DRB6* and *HLA-DRB9*, whereas *HLA-DRB1* alleles **08* and *1415 are linked to the presence of HLA-DRB9. Coexpression of the *HLA-DRB1* locus with the expressed *-DRB3*, *-DRB4*, or *-DRB5* loci frequently occurs. The presence of variable combinations of multiple *HLA-DRB* loci greatly increases the complexity of genotyping.

The basis for *HLA* polymorphism resides in stable inherited variations of the genomic DNA sequence present in the coding regions of the *HLA* genes. Polymorphisms are clustered into discrete hypervariable regions directing changes in the amino acid sequence of the protein ***(6,7)***. Although most *HLA* alleles occur in all ethnic groups, they vary in frequency among these populations. Numerous ambiguities occur when genotyping *HLA* alleles by sequence based-typing. This is primarily because of the presence of *cis/trans* combinations of alleles that occur in certain allelic pairs in heterozygous individuals, such as *HLA-DRB1*0108+*0406* and *HLA-DRB1*010101+*040101* ***(8)***. As a result of the method's intrinsic sequencing by synthesis approach, Pyrosequencing yields out-of-phase sequence information from individual alleles. Out-of-phase Pyrosequencing allows many of the *cis/trans* pairs to be resolved into their individual haplotypes. Thus, providing high-quality data, focused genotyping results, as well as improved *HLA* resolution compared with sequence-based typing (SBT) strategies ***(8–11)***.

2. Materials

2.1. Collection and Purification of DNA From Whole Blood

1. Vacutainer tubes with EDTA anticoagulant may be purchased from Becton-Dickinson, Inc. (Franklin Lakes, NJ). A trained phlebotomist should be available to perform blood draws from volunteers in a manner approved by the local Institutional Review Board.

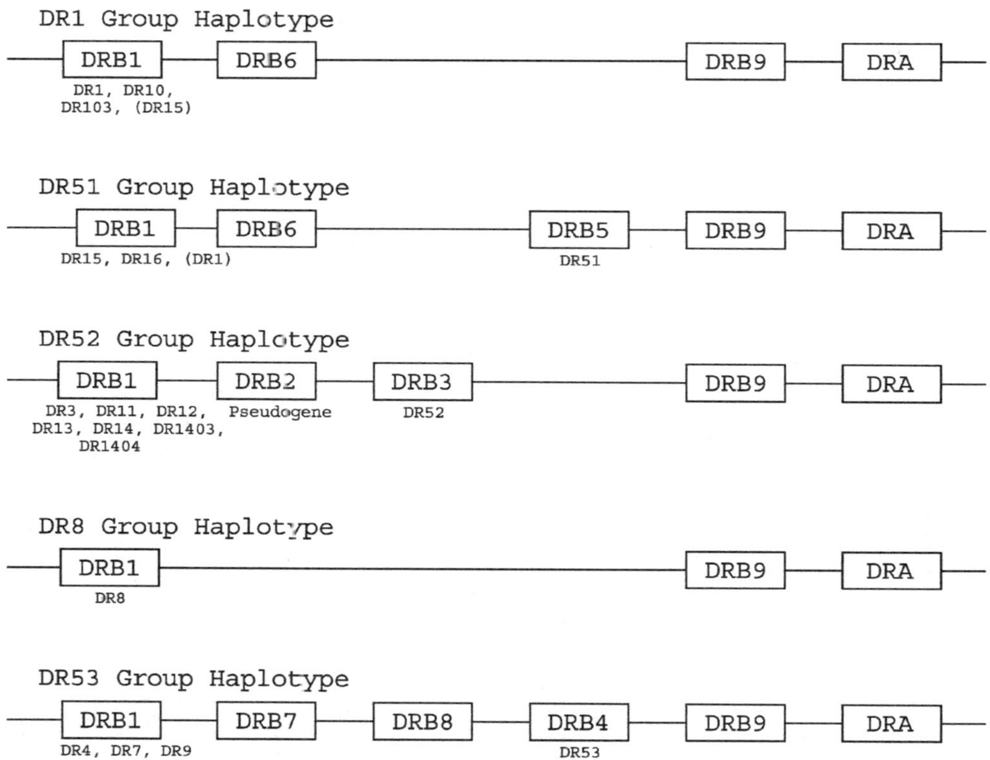

Fig. 1. Multiple human leukocyte antigen *(HLA)-DRB* loci are often present on chromosome 6. *HLA-DRB1* encodes the most frequently occurring expressed *HLA-DRB* loci. Other expressed loci are linked to the various *HLA-DRB1* allelic subgroups. For example, *HLA-DRB3*, *-DRB4*, and *-DRB5* encode the three molecules (DR52, DR53, and DR51, respectively) and are always present with certain *HLA-DRB1* alleles such as DR3, DR4, and DR1, respectively. The loci *HLA-DRB2*, *-DRB6*, *-DRB7*, *-DRB8*, and *-DRB9* represent pseudogenes and are present in different combinations with *HLA-DRB1* allelic subgroups ***(6)***.

2. QIAamp DNA Blood Mini Kit (Qiagen, Inc., Valencia, CA) is used to extract DNA from whole blood. This kit may also be used for preparing DNA from purified cells.
3. Dry blood spots are prepared by spotting whole blood onto S&S 903 Paper (Schleicher and Schuell Inc., Keene, NH).
4. Biotech-grade ethanol and methanol are purchased from Fisher Scientific (Pittsburgh, PA).
5. Punches 1.5 mm in diameter can be obtained using a hand-operated paper puncher or a Wallac Delfia DBS Puncher (Perkin-Elmer Life Sciences Inc., Turku, Finland).

2.2. Estimation of DNA Concentration

1. PicoGreen dsDNA Quantification Kit is obtained from Molecular Probes Inc. (Eugene, OR). The fluorescence-based assay will provide precise measurements of DNA concentration from whole blood-isolated genomic DNA.
2. 10X TE buffer (pH 8.0): dissolve 12.1 g Tris-base and 3.72 g disodium EDTA in 1 L of distilled water. Adjust pH with concentrated HCl to pH 8.0 (*see* **Note 1**).

2.3. PCR

1. 10X *Taq* buffer: 500 m*M* KCl, 100 m*M* Tris-HCl pH 8.3, and 0.01% (w/v) gelatin can be obtained from Applied Biosystems (Foster City, CA). Store frozen at –20°C.
2. Magnesium chloride solution: 25 m*M* solution is prepared by dissolving 0.5 g $MgCl_2$-hexahydrate in 100 mL distilled water. The solution should be transferred into 1-mL aliquots and stored frozen at –20°C.
3. Deoxynucleotide triphosphate solution: 10 m*M* dNTP mixture of dATP, dCTP, dGTP, and TTP can be purchased from Applied Biosystems. Store frozen at –20°C.
4. *Taq* DNA polymerase (5 U/μL) can be purchased from Applied Biosystems. Store frozen at –20°C.
5. DNA oligonucleotide primers for PCR can be purchased from Integrated DNA Technologies Inc. (Ames, IA). Desalted primers should be dissolved in 1X TE buffer pH 8.0 at 10 μ*M* concentration. Store in 1X TE buffer (pH 8.0) at 4°C.
6. Biotinylated DNA oligonucleotide primers containing a 5'-end biotin-tetraethylene glycol modification can be purchased from Integrated DNA Technologies, Inc. (Ames, IA). HPLC-purified primers should be dissolved in 1X TE buffer (pH 8.0) at a concentration of 10 μ*M*. Store in 1X TE buffer (pH 8.0) at 4°C.

2.4. Agarose Gel Electrophoresis

1. Agarose (low electoendosmosis) can be obtained from Roche Inc. (Indianapolis, IN).
2. 5X TBE buffer (1 L): 54 g Tris-base, 27.5 g boric acid, and 3.72 g disodium EDTA in 1 L H_2O. 5X TBE may be stored at room temperature for several months.

3. Ethidium bromide solution (10 mg/mL) can be purchased from Sigma Chemical Company (Saint Louis, MO) (*see* **Note 2**).

2.5. Pyrosequencing-Based Typing

1. Binding buffer: 10 m*M* Tris-HCl, pH 7.6, 2 *M* NaCl, 1 m*M* EDTA, and 0.1% Tween-20. This solution is prepared by mixing 1.2 g Tris-base, 116.9 g NaCl, and 0.34 g disodium EDTA in 900 mL distilled water. Adjust the solution to pH 7.6 with concentrated HCl. Prepare the final solution by addition of 1 mL Tween-20 and adjust the volume to 1 L with distilled water. Store the solution at room temperature.
2. Denaturation solution: 0.2 *M* NaOH. Prepared by dissolving 8 g NaOH in 1 L distilled water. Store at room temperature.
3. Washing buffer: 10 m*M* Tris-acetate, pH 7.6, and 4 *M* glacial acetic acid. The solution is prepared by 1.2 g Tris-base in 833 mL distilled water and 167 mL glacial acetic acid. Store the washing buffer at room temperature.
4. Annealing buffer: 20 m*M* Tris-acetate, pH 7.6 and 2 m*M* magnesium acetate. Prepare by dissolving 2.4 g Tris-base and 0.42 g magnesium acetate-tetrahydrate in 900 mL distilled water. Adjust pH to 7.6 with glacial acetic acid. Add distilled water to adjust the final volume to 1 L. Store at room temperature.
5. *Escherichia coli* single-stranded DNA binding protein can be obtained from Promega Corporation (Madison, WI). Store frozen at –20°C.
6. 96-Well tray formatted MultiScreen 0.45-µm hydrophilic low protein-binding durapore membrane are purchased from Millipore Corp. (Billerica, MA).
7. Pyrosequencing reagent kit and Pyrosequencing 96MA instrument may be purchased from Biotage Inc. (Foxboro, MA).

3. Methods

Advances in sequencing methodology have been applied to SBT of *HLA* loci. Pyrosequencing is advantageous for genetic typing of complex polymorphic motifs within these loci ***(8–11)***. Pyrosequencing-based typing (PSBT) strategies have been designed that exploit the order of nucleotide dispensations to provide DNA sequence for distinguishing nearby *cis/trans* polymorphism combinations. These are the principal allelic pairs responsible for ambiguous genotypes during SBT approaches. For example, of the 1981 possible allelic combinations at the *HLA-DQB1* locus there are 64 (3%) that are ambiguous by SBT and of the 77,815 possible allelic combinations at the *HLA-DRB1* locus there are 3353 (4%) allelic pairs that remain ambiguous. Efficient resolution of *cis/trans* ambiguities within *HLA* loci continues to be an active area of research. Resolution of ambiguous allelic combinations can be realized by cloning and sequencing of isolated DNA, as well as through the development of other advanced sequence-based strategies. Pyrosequencing methods show progress in *HLA* analysis and the method is capable of resolving many of the *cis/trans* combinations ***(8–11)***.

3.1. Collection and Storage of Whole Blood Obtained by Venous Puncture From Human Subjects

1. Collection of whole blood by vein puncture. Sterile vein puncture can be used as a method of obtaining whole blood suitable for DNA extraction and Pyrosequencing. This step should be preformed by a trained phlebotomist on volunteers who have been recruited in a manner specified by the local Institutional Review Board.
2. Whole blood (at least 5 mL) can be drawn into vacutainer tubes containing EDTA as an anticoagulant (Becton-Dickinson Inc.).
3. Samples can be processed immediately or transferred to cryotubes and stored frozen at –80°C. When transferring blood samples biohazard safety precautions should be observed, such as working within a biohazard laminar flow hood, wearing latex gloves, and eye protection. Disposal of used laboratory materials should follow the local regulations of the institution in which the work is being performed.

3.2. Collection and Storage of Dry Blood Spot From Human Subjects

1. An alternative approach to collection of a human sample is to transfer whole blood directly to filter paper to generate dry blood spots for storage and DNA purification. This process uses a sterile needle to deliver a finger prick in order to induce a minimum amount of bleeding, 20–50 μL is sufficient. As when whole blood is collected by vein puncture collection should be performed in a manner specified by the local Institutional Review Board.
2. Whole blood can be transferred directly by pressing the finger onto S&S 903 paper (Schleicher and Schuell).
3. The blood spotted S&S 903 paper is allowed to dry overnight in a biosafety hood.
4. Dry blood spots are sealed from moisture and stored at room temperature. A ziplock plastic bag is sufficient for this step. The sealed samples can be stored in no. 10-sized mailing envelopes. This is a convenient method to separate and avoid contamination of samples collected from different individuals.

3.3. Purification of DNA From Whole Blood Collected by Venous Puncture

1. DNA can be extracted from fresh as well as frozen whole blood. QIAamp DNA Mini Kit (Qiagen, Inc.) has been used for DNA isolation. The method typically yields greater than 0.5 μg of purified genomic DNA per 75 μL of whole blood. The DNA isolation protocol is modified from the QIAamp DNA Blood Mini Kit Handbook provided by Qiagen, Inc. for use with whole blood collected with EDTA as an anticoagulant.
2. The following items are recommended precautions before starting the DNA isolation protocol:

   a. Equilibrate whole blood samples to room temperature before starting.
   b. Prepare a 56°C heating block.
   c. Equilibrate buffer AE to room temperature.

   d. Prepare buffers AW1, AW2, and QIAGEN Protease according to the instructions in the QIAamp DNA Mini Kit Handbook.
   e. Adjust the centrifuge instrument to operate at room temperature.
3. In a 1.5-mL centrifuge tube, thoroughly mix 20 μL QIAGEN Protease with 75–200 μL whole blood. If less than 200 μL whole blood is used, add an appropriate volume of phosphate-buffered saline solution.
4. Add 200 μL buffer AL and vortex for 15 s. Repeat this step for a total of three times.
5. Incubate the mixture for 10 min at 56°C followed by a brief centrifugation to remove droplets from the inside of the lid.
6. Add to the protease-treated blood sample 200 μL of absolute ethanol and mix by vortexing.
7. Transfer the solution to a QIAamp Mini column placed in 2-mL collection tubes. Seal the tubes and centrifuge at 6000*g* for 1 min.
8. Transfer the sample captured on the QIAamp Mini column to a new 2-mL collection tube. Load any remaining solution from **step 4** onto the column. Seal the tube and centrifuge at 6000*g* for 1 min.
9. Add 500 μL buffer AW1 to the column. Seal the column and centrifuge at 6000*g* for 1 min. Place the column into a clean 2-mL collection tube.
10. Add 500 μL buffer AW2 to the QIAamp Mini column. Seal the column and centrifuge at 20,000*g* for 3 min. Place in a new collection tube and centrifuge at 20,000*g* for 1 min.
11. Transfer the QiAamp Mini column to a new 1.5-mL centrifuge tube. Add 50 μL of distilled water by transferring directly onto the center of the membrane of the column. Seal the column and incubate at room temperature for 5 min. Centrifuge at 6000*g* for 1 min. Repeat this step.
12. Store samples either refrigerated or frozen at –20°C. Prior to use in PCR amplification the concentration of the DNA samples should be determined. This will enable the laboratory to optimize the amount of DNA template required to drive the PCR step and allow careful management of samples that human subjects provide only once.

3.4. Isolation of DNA From Dried Blood Spots

1. Purification of genomic DNA from dried blood spot samples follows the protocol outlined by the Qiagen QIAamp DNA Mini Kit and QIAamp DNA Blood Mini Kit Handbook. Multiple 3.2-mm diameter punches can be prepared from each dried blood spot. This can be accomplished by using a hand-operated paper puncher or a motor-operated Wallac Delfia DBS puncher (Perkin-Elmer Life Sciences, Inc.). The hand-operated paper puncher and Wallac Delfia DBS puncher are equipped with a 3.2-mm punch head. Specimens can be punched directly into 96-well trays or sample collection tubes. Six 3.2-mm diameter punches per tube are sufficient for each extraction.
2. The following items are recommended precautions before starting the DNA isolation protocol:

a. Prepare 85, 70, and 56°C heating blocks.
b. Equilibrate buffers AW1 and AW2 to room temperature.
c. Adjust the centrifuge instrument to operate at room temperature.

3. Using a 3.2-mm diameter punch head, transfer six punches of the dried blood spot to a 1.5-mL centrifuge tube. Add 180 μL buffer ATL.
4. Place the tube in an 85°C heating block for 10 min. Following the incubation briefly centrifuge to remove droplets from the lid.
5. Add 20 μL proteinase K and mix thoroughly. Incubate the sample at 70°C for 10 min. Briefly centrifuge to remove droplets from the inside of the lid.
6. Add 200 μL absolute ethanol. Mix thoroughly and centrifuge to remove droplets from the lid.
7. Transfer the sample to the QIAamp Spin Column and place the column into a 2-mL collection tube. Seal the column and centrifuge at 6000*g* for 1 min. Transfer the column to a clean 2-mL collection tube and discard the filtrate.
8. Open the column and add 500 μL buffer AW1 to the center of the membrane. Seal the column and centrifuge at 6000*g* for 1 min. Transfer the column to a clean 2-mL collection tube and discard the filtrate.
9. Open the column and add 500 μL buffer AW2 to the center of the membrane. Seal the column and centrifuge at 20,000*g* for 3 min.
10. Transfer the column to a clean 1.5-mL centrifuge tube. Open the column and transfer 100 μL buffer AE to the center of the membrane. Elute the DNA sample from the column and centrifuge at 6000*g* for 1 min. Repeat this step one time and discard the column.
11. Store samples either refrigerated or frozen at –20°C. Prior to use in PCR amplification the concentration of the DNA samples should be determined. This will enable the laboratory to optimize the amount of DNA template required to drive the PCR step and allow management of samples that human subjects provide only once.

3.5. Determining the Concentration of Purified DNA

1. DNA concentration can be determined using the PicoGreen dsDNA Quantitation Kit (Molecular Probes). The method allows estimation of DNA concentration by comparison of the fluorescent signal obtained from each sample with that collected using a dilution series of a DNA standard. Signal can be measured with a fluorescent microplate reader using excitation wavelength 484 nm, emission wavelength 538 nm, and a 530-nm bandpass filter. The DNA quantification protocol is modified from the method provided by Molecular Probes, Inc. for use in a 96-well tray.
2. The following items compose the recommended precautions to be observed before starting the DNA concentration protocol:
   a. Allow the PicoGreen reagents, dimethyl sulfoxide stock solution, and DNA samples to warm to room temperature.
   b. Prepare TE buffer (10 m*M* Tris-HCl, 1 m*M* EDTA, pH 7.5) for use in diluting DNA samples and assay reagents.

c. Use plastic rather than glass containers to prepare samples and reactions to avoid possible absorption of the fluorescent dye to the glass surfaces.
d. Protect the solutions from exposure to light as the PicoGreen reagent is susceptible to photobleaching.
e. Use all solutions within a few hours of their preparation.

3. Freshly prepare an aqueous working solution of the PicoGreen reagent by preparing a 200-fold dilution of the concentrated dimethyl sulfoxide solution in 1X TE, pH 8.0. For 20 assays mix 0.1 mL PicoGreen dsDNA quantitation reagent with 19.9 mL 1X TE, pH 8.0.
4. Prepare two dilutions of λ DNA. Using the 0.1 mg/mL λ DNA standard provided with the PicoGreen dsDNA quantitaiton kit, the high-range standard stock solution is prepared by mixing 30 μL λ DNA with 1.47 mL 1X TE, pH 7.5. The low-range standard stock solution is prepared by mixing 3 μL of the high-range standard with 1.5 mL 1X TE, pH 7.5.
5. The high-range standard DNA curve for a range of 1 ng/mL to 1 μg/mL can be prepared from the 2 μg/mL high-range DNA standard stock solution as below:

Concentration	1X TE buffer	High-range DNA standard (2 mg/mL)	Diluted PicoGreen reagent
1 μg/mL	—	1 mL	1 mL
100 ng/mL	900 μL	100 μL	1 mL
10 ng/mL	990 μL	10 μL	1 mL
1 ng/mL	999 μL	1 μL	1 mL
No DNA control	1 mL	—	1 mL

6. The low-range standard DNA curve for a range of 25 pg/mL to 25 ng/mL can be prepared from the 50 ng/mL low-range DNA standard stock solution as next:

Concentration	1X TE buffer	Low-range DNA standard (50 ng/mL)	Diluted PicoGreen reagent
25 ng/mL	—	1 mL	1 mL
2.5 ng/mL	900 μL	100 μL	1 mL
250 pg/mL	990 μL	10 μL	1 mL
25 pg/mL	999 μL	1 μL	1 mL
No DNA control	1 mL	—	1 mL

7. Preparation of the purified DNA samples for quantification is performed using each of the purified DNA samples as next:

Dilution factor	1X TE buffer	DNA sample	Diluted PicoGreen reagent
1:2	50 μL	50 μL	100 μL
1:20	90 μL	10 μL	100 μL
1:200	99 μL	1 μL	100 μL

8. DNA prepared for the standard curves and the dilutions of sample DNA are mixed thoroughly with an equal volume of diluted PicoGreen reagent and 0.1 mL of each is transferred to a 96-well microplate suitable for use in a fluorescence plate reader. Incubate these reactions for 2–5 min at room temperature.
9. Fluorescent signals are measured in a fluorescent microplate reader. The excitation wavelength is 484 nm, emission wavelength is 538 nm, and the instrument should use a 530-nm bandpass filter.
10. Linear regression can be used to analyze the background-corrected signals. Concentration of the sample DNAs are estimated from the relationship of concentration and signal obtained from the standard curves. If none of the sample signals correspond to the signal range measured for the standard then appropriate fresh dilutions of the sample should be prepared. In the event that more than one of the sample dilutions is within range of the standard curves, their estimated concentrations may be reported as the mean concentration.
11. Adjust the concentration of the samples to 5 ng/μL using distilled water. Store diluted DNA samples in cryotubes frozen at –20°C.

3.6. Polymerase Chain Reaction

1. Amplification of *HLA* loci follows the protocol described by Ringquist et al. *(9)* and Ringquist et al. *(8)*. The reactions are performed in 50-μL volumes using the PCR forward and biotinylated reverse primers indicated in **Table 1**.
2. The following solutions should be prepared for use during PCR amplification of *HLA* loci and can be stored at –20°C until needed (*see* **Note 3**):
   a. 10X *Taq* buffer: 500 m*M* KCl, 100 m*M* Tris-HCl, pH 8.3, 0.01% (w/v) gelatin.
   b. Magnesium chloride solution: 25 m*M* $MgCl_2$.
   c. Deoxynucleotide triphosphate solution: 10 m*M* dNTP mixture.
   d. Forward primer: 10 μ*M* forward PCR primer in TE buffer, pH 8.0.
   e. Biotinylated primer: 10 μ*M* biotinylated reverse PCR primer in TE buffer, pH 8.0.
3. Thaw reagents in an ice-water bath or by gentle warming to room temperature. Excessive incubation at warmer temperatures may reduce the shelf life of the reagents. Mixtures are prepared using DNA purified from whole blood or extracted from dried blood spots:

Solution	DNA purified from whole blood	DNA purified from dried blood spot
10X *Taq* buffer	5 μL	5 μL
25 m*M* $MgCl_2$	4 μL	4 μL
10 m*M* dNTP	1 μL	1 μL
10 μ*M* forward primer	1 μL	1 μL
10 μ*M* biotinylated primer	1 μL	1 μL
Distilled water	36 μL	18 μL
5 U/μL *Taq* DNA polymerase	0.2 μL	0.2 μL
5 ng/μL DNA sample	2 μL	—
Dry blood spot extract	—	20 μL

Table 1
Oligonucleotides Primers for PCR Amplification of Selected HLA Loci

Name	Sequence	Product length	Comment
HLA-DQB exon 2			
SR25	TTTGACCCCGCAGAGGATTTCGTG	271 bp	*DQB1* forward primer use with SR22 or SR24
SR22	Biotin-TEG-CTCTCCTCTGACRGATCCC		*DQB1* reverse primer, alleles 05/06 use with SR25
SR24	Biotin-TEG-CTCGCCGCTGCAAGGTCGT		*DQB1* reverse primer, alleles 02/03/04 use with SR25
HLA-DRB exon 2			
SR88	AATCCCCACAGCACGTTTCCTG	297 bp	*DRB*ampA(C) forward primer use with SR81
SR89	AATCCCCACAGCACGTTTCTTG	297 bp	*DRB*ampA(T) forward primer use with SR81
SR147	GTTTCTTGGAGCAGGTTAAAC	253 bp	*DRB1* forward primer for groups DR4 use with
SR81			
SR81	Biotin-CCGCTGCACTGTGAAGCTCT		*DRB*ampB biotinylated reverse primer

PCR primers were designed following the recommendations of the 11th International Histocompatibility Workshop ***(17)***. The forward PCR primers were modified based on the constraint of avoiding formation of self-priming 3' secondary structure during pyrosequencing (*see* **Notes 9** and **10**).

4. Thermal cycling conditions are 96°C incubation for 3 min followed by 32 cycles at 96, 55, and 72°C incubated for 30 s at each step. It is recommended that thermal cycling be followed by a final incubation for 5 min at 72°C. Samples can be stored at 4°C or frozen prior to Pyrosequencing (*see* **Note 4**).

3.7. Viewing the Results of PCR Amplification by 1.5% Agarose Gel Electrophoresis

1. Quality control of PCR amplification of *HLA* loci can be accessed by gel electrophoresis using 1.5% agarose gel electrophoresis. The following protocol for agarose gel electrophoresis corresponds to that described by Sambrook et al. *(12)*. Begin by sealing the edges of a clean electrophoresis tray with laboratory tape. Place the sealed gel tray onto a level surface.
2. Prepare a mixture of 1X TBE and powdered agarose to make a sufficient volume of gel solution to fill the tray to a level of 50–100 mm. For example, for a preparation of 1.5% agarose in 100 mL 1X TBE use 1.5 g powered agarose. The 1X TBE buffer is 0.9 *M* Tris-HCl pH 8.0, 0.9 *M* boric acid, and 0.002 m*M* EDTA. A 5X solution of TBE can be prepared ahead of time, stored at room temperature, and diluted with distilled water prior to use.
3. Add a magnetic stir bar and heat the slurry in an Erlenmeyer flask, loosely sealed with aluminum foil, using an electric stirrer–hot plate. Make certain that the agarose has completely dissolved. It is common that for agarose gel solutions containing greater than 1.2% agarose that this process may take a few minutes of constant boiling.
4. A microwave oven may also be used to dissolve the agarose. Caution should be observed when handling the heated agarose solution in that it can become superheated and create a safety hazard by unexpectedly boiling over when the container is handled. When using the microwave oven the aluminum foil and magnetic stir bar should be omitted.
5. The appearance of undissolved agarose in the mixture can be determined by observing the presence of small clumps of semitransparent material when the mixture is stirred. This material must be completely dissolved by heating prior to using it to prepare a gel.
6. The dissolved agarose mixture is cooled to 60°C and ethidium bromide is added to a final concentration 0.5 μg/mL. Thorough mixing of the ethidium bromide into the agarose solution is required.
7. The agarose–ethidium bromide solution is poured into the sealed gel tray. Air bubbles can be removed from the gel by a Pasteur pipet. The gel comb is added to the mixture and the agarose is allowed to solidify upon cooling.
8. After the agarose has solidified remove the gel comb and the sealing tape. The gel tray can be placed in the electrophoresis chamber. Add 1X TBE to the electrophoresis chamber to cover the gel.
9. DNA samples can be mixed with gel-loading buffer. 6X loading buffer 0.25% bromophenol blue, 0.25% xylene cyanol FF, and 30% (v/v) glycerol is recommended but other equally appropriate loading buffers are available *(12)*. PCR

amplification should generate enough DNA so that 5 μL of the amplicon mixed with 1 μL of 6X gel-loading solution should be sufficient to visualize the product. All of the mixture is loaded.

10. DNA-size standards for agarose gel electrophoresis may be purchased from commercial vendors. These are available in a number of variations. PCR products to be examined using Pyrosequencing are typically less than 500 bp in length. This makes the 100-bp ladder standards convenient for the purpose of evaluating the product.
11. After loading the size standard and samples onto the gel the lid of the electorphoresis tank should be closed and the electrical leads connected. DNA is negatively charged and will migrate toward the anode (red lead). Apply a voltage of between 1 and 5 V/cm. For a 20-cm gel box this would be between 20 and 100 V.
12. Once the gel has run sufficiently the lead dye should have migrated roughly 3 cm into the agarose. Turn off the electric current and remove the gel tray from the tank. Be careful not to allow the agarose gel to slip off of the tray as it will likely tear. The DNA bands may be visualized by ultraviolet light and the data recorded by photography or captured by video imaging.
13. Confirmation of successful PCR amplification is determined when a product of the expected length is observed. Products of other lengths or multiple products are considered evidence that the PCR amplification did not provide the proper material to prepare template for Pyrosequencing.

3.8. Pyrosequencing of HLA *Loci*

1. HLA Pyrosequencing can be performed on a PSQ 96MA or PSQ 96HA system. Long-read Pyrosequencing assays described here were performed on the PSQ 96MA. Pyrosequencing reactions are performed using reagents provided with the PSQ 96MA Sample Preparation Kit and PSQ 96 SQA Reagent Kit purchased from Biotage, LLC. **Figure 2** illustrates a Pyrogram for *HLA-DQB1*0201-02+*0203* obtained using DNA purified from a heterozygous individual (*see* **Note 5**).
2. Materials to prepare prior to the Pyrosequencing protocol:
   a. Binding buffer: 10 m*M* Tris-HCl pH 7.6, 2 *M* NaCl, 1 m*M* EDTA, and 0.1% Tween-20.
   b. Annealing buffer: 20 m*M* Tris-acetate pH 7.6 and 2 m*M* magnesium acetate.
   c. Denaturation solution: 0.2 *M* NaOH.
   d. Washing buffer: 10 m*M* Tris-acetate, pH 7.6 and 4 *M* glacial acetic acid.
3. Samples are prepared using 10 μL of amplified DNA from the PCR mixtures but as much as 46 μL of PCR product can be used. These are mixed with 4 μL of streptavidin-coated beads purchased from Amersham-Pharmacia Biotech, Inc. and the volume is adjusted to 50 μL with binding buffer.
4. Place the mixture of PCR product and streptavidin-coated beads onto an Eppendorf Thermal Mixer and vortex at room temperature using 1400 rpm for 10 min.
5. Transfer the streptavidin-coated beads to a Millipore 0.45-μm membrane and collect the biotinylated PCR product–streptavidin-coated bead complex by vacuum filtration.

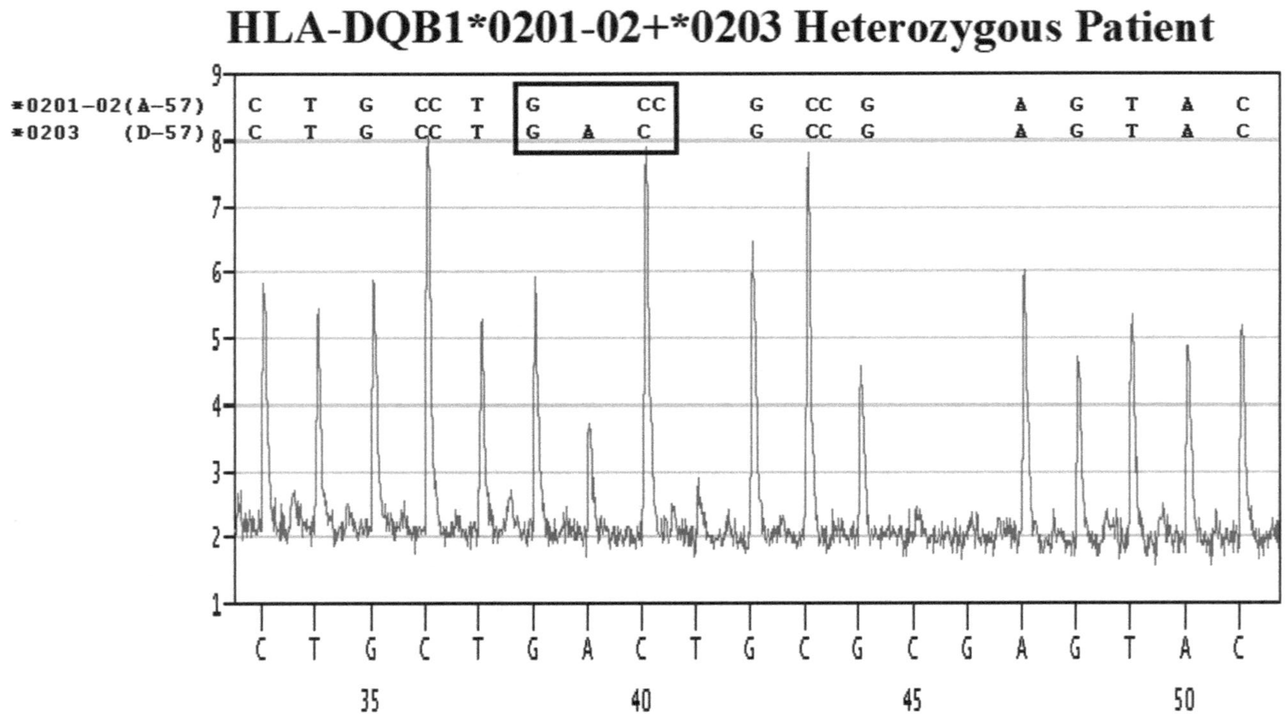

Fig. 2. Pyrogram® of human leukocyte antigen *(HLA)-DQB1* alleles **0201-02+*0203* obtained from DNA isolated from a *HLA* heterozygous individual. DNA was prepared from whole blood and PCR amplified using *HLA-DQB1*-specific primer SR25 and SR24. The sequencing primer was PSQ3 and the Pyrosequencing step used the nucleotide dispensation order indicated in **Table 2**. The sequence corresponding to nucleotide dispensations 33 through 51 are indicated for the **0201-*02* and **0203* alleles. The boxed residues indicate the position of codon 57 triplet corresponding to alanine or aspartate (*see* **Note 8**).

6. Beads are resuspended in 150 µL denaturation buffer and incubated for 2 min at room temperature. Denaturation buffer is removed by vacuum filtration.
7. Immediately after removal of the denaturation buffer, and while still under vacuum, the pH of the PCR product–bead complex is neutralized by addition of 200 µL washing buffer. Repeat this step for a total of two washings.
8. Resuspend beads in 50 µL annealing buffer and transfer 45 µL to the Pyrosequencing tray.
9. Add the appropriate Pyrosequencing primer to each well in a volume of 5 µL using 3 µ*M* stock solution of primer. Recommended Pyrosequencing primers are listed in **Table 2**. Annealing was performed by incubation at 80°C for 2 min prior to Pyrosequencing. Single-stranded binding protein is added at a final concentration of 4 µg per Pyrosequencing reaction (*see* **Notes 6** and **7**).

4. Notes

1. Preparation of aqueous solutions. All solutions are prepared in deionized water (18.2 mOhm-cm) filtered free of organic content.
2. Disposal of ethidium bromide. Ethidium bromide is a carcinogenic compound and should be handled with appropriate care. Dilute solutions of ethidium bromide should be decontaminated before disposal. For 0.5 µg/mL solutions of ethidium bromide the method described by Lunn and Sansone ***(13)*** and an alternate method described by Bensaude ***(14)*** are recommended. Treatment of dilute solution of ethidium bromide with bleach (hypochlorite) is not recommended. The latter method has been shown to convert the dye into a compound that is mutagenic ***(12,15)***.
3. dUTP and uracil-DNA glycosylase can be used to reduce sample contamination ***(16)***. Contamination of PCR amplifications from previous PCR amplicons is a frequent problem in laboratories in which the same PCR primers are used repetitively. Incorporation of dUTP and use of uracil DNA glycosylase has been shown to eliminate the carryover of product. The protocol is compatible with Pyrosequencing in that the dU-containing PCR product serves as an efficient template for sequencing.
4. Validation of PCR primers by gradient thermal cycler amplification. Optimization of PCR conditions is conveniently performed using a temperature gradient capable thermal cycler. This instrument can be purchased from the MJ Research division of Bio-Rad Laboratories Inc. A range of PCR-annealing temperatures can be examined and the relative yield of PCR product determined by comparing the intensity of DNA bands after ethidium bromide staining of bands separated by 1.5% agarose gel electrophoresis.
5. Recommended DNA standards for use during Pyrosequencing of *HLA* samples. Human lymphoblastoid cell lines of known *HLA* identity can be obtained from the International Histocompatibility Working Group Cell and Gene Bank. These samples have been genotyped for *HLA* class I and class II loci ***(17)***.
6. Validation of Pyrosequencing primers. Pyrosequencing reactions can give background signals from self-priming of the 3'-end of the template strand or by self-annealing of the sequencing primer. The validation of the Pyrosequencing step

Table 2
Oligonucleotides Primers for PSBT of Selected *HLA* Genotypes

Name	Sequence	Sequencing primer annealing site	Recommended nucleotide dispensation order (*see* **Note 11**)
Pyrosequencing primer for HLA-DQB1 exon 2			
SR26	GACCCCGCAGAGGATTTCG	Residues –10 to 9	TGCTCACAGTTAGGCATGTGCTACTCACACTGGACAGAGCG TCGTGCGGTCATGTGAGCAGATCAGCAT
SR27	CAACGGGACGGAGCGCGT	Residues 41 to 58	GCGGTCATGTGAGCAGACTGACATCTATACGAGAGAGATA CGTCGACGCTCGACAGCGACGTGGATGTGTATC
PSQ1	ACGGAGCGCGTGCG	Residues 48 to 61	TACGGTTGTGAGCCAGAACTAGCATCTATACGAGA
PSQ3	TCGACAGCGACGTG	Residues 106 to 119	GGAGTGTATCGGCTAGCGTGACCGCTGCATGGCTG CTGACTGCGCGAGTACTGACTAGCAGAGACGATC
SR30	GCCGAGTACTGGAACAGCC	Residues 159 to 177	GAGACGATCTGAGGAGAACCGGAGCGTCAGTGACACGGAG TGATGCAGACACACTACGAGTGGCGAGCTACGC
Pyrosequencing primers for HLA-DRB exon 2			
SR88	AATCCCCACAGCACGTTTCCTG	Residues –11 to 11	GCAGTGCACTGTGCATAGTCAGTCGATGAGTGTCTCATAC TCTGCACTGACTAGAGCGATGATAGCTAG
SR89	AATCCCCACAGCACGTTTCTTG	Residues –11 to 11	Same as for SR88
SR83	TCAATGGGAC	Residues 40 to 49	GAGCATGCAGATACATGCATGTAGTCTGCTGAGCAGACTAG CAGCATACGATCTATATCACAGAGAGTATGCTGC
SR84	TCAATGGGAT	Residues 40 to 49	Same as for SR83
SR85	TCAACGGGAC	Residues 40 to 49	Same as for SR83
SR86	GGGCGGCCT	Residues 147 to 155	GAGTGACTGCTGAGCTACTGACATGACAGCAGAGCATCTG AGCAGCACATCAGATCGCGC
SR87	GGAACAGCCA	Residues 169 to 178	GAGACATCTGCTGCTGAGCACAGAGATCGTGACATGACTG TCGTGCACGTAGACTACTGTGTACTGCAGACT
SR131	TGGAGCAGGTTA	Residues 10 to 21; *DRB1*04* specific in *DRB1*07*, *DRB4* background	GACTGACTGACTGACTGACTGACTGACTGACTGAC

can be performed in order to examine the signals because of these events. A set of negative control reactions should be examined to estimate the background Pyrosequencing signal. A recommended set of negative control samples are listed next:

a. Negative control 1: biotinylated PCR primer and sequencing primer in the Pyrosequencing reaction.
b. Negative control 2: biotinylated PCR primer alone in the Pyrosequencing reaction.
c. Negative control 3: sequencing primer alone in the Pyrosequencing reaction.
d. Negative control 4: Pyrosequencing PCR negative control.
e. Negative control 5: Pyrosequencing PCR negative control without sequencing primer.

7. Extending the Pyrosequencing read length. Pyrosequencing has enabled DNA sequence reads of up to 150 residues ***(8,9,18,19)***. Optimization of the Pyrosequencing reaction conditions has been reported to improve read length ***(19,20)*** and focuses primarily on the use of enantiomer pure 2'-deoxyadenosine-5'-*O*'-(1-thiotriphosphate) and the use of single-stranded DNA binding protein ***(18,19)***.
8. Exploiting out-of-phase Pyrosequencing for *cis/trans* ambiguities. An advantage of Pyrosequencing is that single nucleotide dispensation results in out-of-phase sequencing of the individual alleles. A result of this is that sequence information can be obtained from each allele independently ***(18)***. Exploitation of out-of-phase Pyrosequencing can allow resolution of allelic combinations that are ambiguous by conventional in-phase sequencing methods ***(8–11)***. Allele-specific *HLA* PCR and Pyrosequencing primers greatly improve the resolution of *HLA* genotyping ***(8)***.
9. Selection of oligonucleotide primers. Primers for PCR and Pyrosequencing have been described for *HLA-DQB1* and *-DRB1* exon 2. Complete sequencing of the exon 2 from these loci is obtained by use of multiple Pyrosequencing primers. A single PCR amplification can provide sufficient template for analysis.
10. Primer design software. Tools for designing primers for PCR and Pyrosequencing are available from a number of sources. Web-based software has been generated specific for Pyrosequencing and is available at http://imgen.ccbb.pitt.edu ***(21,22)***. The software accepts the name of the genetic loci under investigation and returns PCR primers and sequencing primer candidates for laboratory investigation. Another convenient software tool is provided by Biotage, LLC and allows the user to enter a text file of the sequence under investigation. The Biotage software also returns a set of suggested primers for laboratory use. Standard *HLA* primers for PCR amplification are listed at the IHWC website at http://ihwc.org. These primers and the recommended PCR conditions are known to provide robust amplification of specific *HLA* loci and *HLA* allele subgroups.
11. Design of nucleotide dispensation order. Optimization of nucleotide dispensation order enables out-of-phase sequencing, leading to resolution of allelic combinations. *Cis/trans* combinations of alleles are the principal cause of ambiguous sequence information when analyzing *HLA* samples. Use of multiple dispensa-

tion orders can be applied to the same *HLA* sample preparation. This can aid in the detection of alleles by producing sets of unique Pyrosequencing signals that can resolve allelic pairs.

Acknowledgments

This work was supported by grants from the National Institutes of Health (NIH), U19-AI056374-01 Autoimmunity Centers of Excellence (S. R. and M. T.), RO1DK24021(M. T.) from the NIH, and ERHS no. 00021010 (S. R. and M. T.) from the Department of Defense.

References

1. Mungall, A. J., Palmer, S. A., Sims, S. K., et al. (2003) The DNA sequence and analysis of human chromosome 6. *Nature* **425,** 805–811.
2. Marsh, S. G. (1999) Nomenclature for factors of the HLA system, update March 1999. WHO Nomenclature Committee for factors of the HLA system. *Hum. Immunol.* **60,** 627.
3. Robinson, J., Waller, M. J., Parham, P., et al. (2003) IMGT/HLA and IMGT/MHC: sequence databases for the study of the major histocompatibility complex. *Nucl. Acids Res.* **31,** 311–314.
4. Stewart, C. A., Horton, R., Allcock, R. J., et al. (2004) Complete MHC haplotype sequencing for common disease gene mapping. *Genome Res.* **14,** 1176–1187.
5. Campbell, R. D. and Trowsdale, J. (1993) Map of the human MHC. *Immunology Today* **14,** 349–352.
6. Ringquist, S., Nichol, L., and Trucco, M. (In press) Transplantation genetics. In: *Emery and Rimoin's Principles and Practice of Medical Genetics 5th ed.,* (Rimoin, D., Connor, J. M., Pyeritz, R., Korf, B., and Emery, A., eds.), Churchill Livingstone.
7. Luppi, P. and Trucco, M. (2005) Type 1 diabetes as an autoimmune disease. In: *Measuring Immunity: Basic Science and Clinical Practice*, (Lotze, M. T. and Thomson, A. W., eds.), Elsevier Academic Press, London, UK, pp.494–504.
8. Ringquist, S., Alexander, A. M., Styche, A., Pecoraro, C., Rudert, W. A., and Trucco, M. (2004) HLA class II DRB high resolution genotyping by Pyrosequencing: comparison of group specific PCR and Pyrosequencing primers. *Hum. Immunol.* **65,** 163–174.
9. Ringquist, S., Alexander, A. M., Rudert, W. A., Styche, A., and Trucco, M. (2002) Pyrosequence based typing of alleles of the HLA-DQB1 gene. *Biotechniques* **33,** 166–175.
10. Ramon, D., Braden, M., Adams, S., Marincola, F. M., and Wang, L. (2003) Pyrosequencing: a one-step method for high resolution HLA typing. *J. Transl. Med.* **1,** 9.
11. Entz, P., Toliat, M. R., Hampe, J., et al. (2005) New strategies for efficient typing of HLA class-II loci DQB1 and DRB1 by using Pyrosequencing. *Tissue Antigens* **65,** 67–80.

12. Sambrook, J., Fritsch, E. F., and Maniatis, T. (1989) *Molecular Cloning: A Laboratory Manual, 2nd ed.*, Cold Spring Harbor Laboratory Press, Cold Spring Harbor, New York.
13. Lunn, G. and Sansone, E. B. (1987) Ethidium bromide: destruction and decontamination of solutions. *Anal. Biochem.* **162,** 453–458.
14. Bensaude, O. (1988) Ethidium bromide and safety-readers suggest alternative solutions. *Trends Genet.* **4,** 89–90.
15. Quillardet, P. and Hofnung, M. (1988) Ethidium bromide and safety-readers suggest alternative solutions. *Trends Genet.* **4,** 89.
16. Longo, M. C., Berninger, M. S., and Hartley, J. L. (1990) Use of uracil DNA glycosylase to control carry-over contamination in polymerase chain reactions. *Gene* **93,** 125–128.
17. Faas, S. J., Menon, R., Braun, E. R., Rudert, W. A., and Trucco, M. (1996) Sequence-specific priming and exonuclease-released fluorescence detection of HLA-DQB1 alleles. *Tissue Antigens* **48,** 97–112.
18. Garcia, C. A., Ahmadian, A., Gharizadeh, B., Lundeberg, J., Ronaghi, M., and Nyren, P. (2000) Mutation detection by Pyrosequencing: sequencing of exons 5-8 of the p53 tumor suppressor gene. *Gene* **253,** 249–257.
19. Gharizadeh, B., Nordstrom, T., Ahmadian, A., Ronaghi, M., and Nyren, P. (2002) Long-read pyrosequencing using pure 2'-deoxyadenosine-5'-O'-(1-thiotriphosphate) Sp-isomer. *Anal. Biochem.* **301,** 82–90.
20. Ronaghi, M. (2000) Improved performance of Pyrosequencing using single-stranded DNA-binding protein. *Anal. Biochem.* **286,** 282–288.
21. Alexander, A. M., Pecoraro, C., Styche, A., et al. (2005) SOP3: a web-based tool for selection of oligonucleotide primers for single nucleotide polymorphism analysis by Pyrosequencing. *Biotechniques* **38,** 87–94.
22. Ringquist, S., Pecoraro, C., Gilchrist, M. S., et al. (2005) SOP[3]v2: web-based selection of oligonucleotide primer trios for genotyping of human and mouse polymorphisms. *Nucl. Acids Res.* **33,** W548–W552.

11

Pyrosequencing® of Phage Display Libraries for the Identification of Cell-Specific Targeting Ligands

Ahad A. Rahim

Summary

High combinatorial phage display libraries have become an important tool in the search for ligand–receptor interactions. The advantage this approach offers is the ability to screen large repertoires of peptides, displayed on the coat proteins of bacteriophages, against a target at the same time. In addition, no prior knowledge is required of the target or the ligand. However, to characterize the peptides of interest a short length of the bacteriophage genome that encodes the peptide sequence requires DNA sequencing. The number of candidate bacteriophages can be large and so sequencing is expensive, time-consuming, and laborious. Therefore, a methodology using Pyrosequencing® has been developed where 96-phage displaying a seven amino acid peptide can be analyzed simultaneously within 45 min and at a fraction of the cost associated with traditional automated Sanger sequencing.

Key Words: Pyrosequencing; phage; peptide.

1. Introduction

Bacteriophages are viruses that infect a variety of Gram-negative bacteria including *Escherichia coli*. In 1985, George Smith showed that fragments of foreign DNA could be inserted into the *pIII* gene of the filamentous phage genome. This resulted in a fusion protein consisting of the protein encoded by the gene linked to the peptide coded for by the inserted oligonucleotide. The *pIII* gene codes for a structural component of the phage surface and the insertion resulted in display of the foreign peptide on the surface of the phage coat *(1)*. This led to the birth of phage display technology and the subsequent production of high-complexity phage display libraries. Phage display involves the expression of peptides or proteins (such as antibodies) on the surface of a phage. A DNA sequence of interest is inserted into a region of the phage genome such

From: *Methods in Molecular Biology, vol. 373: Pyrosequencing® Protocols*
Edited by: S. Marsh © Humana Press Inc., Totowa, NJ

that the encoded peptide would be expressed or "displayed" as protein fusion on the surface of filamentous phage as demonstrated by Smith et al. *(1)*. The advantage this technology offers is that large complex phage display libraries can be produced in which several billion variants can be constructed all at once. This is far preferable to genetically engineering proteins or peptide variants that must be expressed, purified, and analyzed individually, which is hugely time consuming and laborious. These high-complexity phage display libraries can then be used to screen and purify specific phage displaying sequences that have increased binding abilities to a particular target of interest by a process termed "biopanning." This process involves phage undergoing repeated screening against the target with rounds of binding and washing to eliminate phage that have no, or weak, interaction with the target. Those phage that are selected on the basis of their stronger interaction with the target are rescued and amplified in bacteria before being isolated and purified either for reapplication to the target under greater stringency or for DNA sequencing as the final step in order to determine the peptide sequence that the phage is displaying. This high-throughput method for screening of a large number of peptides against a target has made it an extremely useful tool that has been used in a number of application and studies including the identification of novel ligands to cellular receptors *(2–6)*. It is therefore an important tool for identifying new targeting moieties for a number of applications including drug delivery and gene therapy vectors.

However, the DNA sequencing of candidate phage after the biopanning process is a major rate-limiting step. Often, large numbers of candidate phage need sequencing, especially if enrichment is low or greater numbers are required to look for patterns in peptide sequences. The cost of analyzing the phage using traditional automated Sanger sequencing can be uneconomical, especially considering that just the short oligonucleotide inserted into the phage genome requires sequencing. In addition, DNA extraction and purification from phage and then Sanger sequencing is a time-consuming and laborious procedure.

Here, an alternative methodology is presented by which Pyrosequencing (Biotage AB, Uppsala, Sweden) is used to DNA sequence the short oligonucleotide of interest in the phage genome derived from the PhD-7 Phage Display Peptide Library (New England Biolabs, Beverly, MA). Pyrosequencing produces DNA sequences for 96 phage simultaneously within 45 min, and at a fraction of the cost when compared with traditional automated Sanger sequencing.

2. Materials

2.1. Cell Culture and Biopanning

1. Dulbecco's modified Eagle's medium (Invitrogen, Paisley, UK) supplemented with 10% fetal bovine serum (Invitrogen).

2. 1% Bovine serum albumin (BSA) (Sigma-Aldrich, St. Louis, MO) made up in phosphate-buffered saline (Invitrogen) and purified through a filter.
3. PhD-7 Phage Display Peptide Library (New England Biolabs) (*see* **Note 1**).
4. Tris-buffered saline (TBS) (Invitrogen).
5. 0.1 and 0.2% Tween-20 made up in TBS (Sigma-Aldrich).
6. ER2738 *E. coli* (New England Biolabs).

2.2. Amplification of Phage Eluate

1. ER2738 *E. coli* (New England Biolabs).
2. Tetracycline (Sigma-Aldrich) made up in 50% (v/v) ethanol to concentrations of 20 and 40 mg/mL.
3. Luria-Bertani (LB) bouillon (Merck, Whitehouse Station, NJ).
4. Tetracycline plates: 15 g agar per liter of LB are mixed, autoclaved, and left to cool a little but not solidify. Tetracycline is added so that the final concentration is 40 μg/mL. Plates are then poured and stored at 4°C.
5. Polyethylene glycol 8000 (PEG) (Sigma-Aldrich)/NaCl: 20% (w/v) PEG and 2.5 *M* NaCl. Filter and store at room temperature.

2.3. Phage Titering

1. ER2738 *E. coli* (New England Biolabs).
2. LB/isopropyl β-D-1-thiogalactopyranoside (IPTG)/5-bromo-4-chloro-3-indolyl-β-D-galactosidase (X-gal) plates: 15 g agar per liter of LB are mixed and autoclaved before being left to cool a little but not solidify. A mixture of IPTG (Sigma-Aldrich) and X-gal (Melford Laboratories, Ipswich, UK) (1.25 g of IPTG and 1 g of X-gal made up in 25 mL of dimethyl formamide) is added and mixed into the LB/agar solution and plates are poured under sterile conditions. The plates can be stored at 4°C in the dark.
3. Agarose top: for 1 L, 10 g of bactotryptone, 5 g yeast extract, 5 g NaCl, 1 g $MgCl_2 6H_2O$, and 7 g of agarose are mixed and autoclaved. Smaller aliquots are made for ease of melting when required using a microwave.

2.4. Amplification of Phage From an Infected Bacterial Colony

1. ER2738 *E.coli* (New England Biolabs).
2. Sterile wooden picks: simple wooden toothpicks can be used, which are sterilized by autoclaving.
3. Glycerol: filter purified.

2.5. Polymerase Chain Reaction

1. Forward primer: 5'-ATTCGCAATTCCTTTAGTGGTA-3' (Sigma-Genosys, Haverhill, UK) (*see* **Note 2**).
2. Reverse primer: 5'-biotin-GGGATTTTGCTAAACAACTTT-3' (Sigma-Genosys) (*see* **Note 3**).
3. PCR Gold buffer (Applied Biosystems, Foster City, CA).
4. $MgCl_2$ (Applied Biosystems).

5. dNTP mix (Invitrogen).
6. Amplitaq Gold (Applied Biosystems).

2.6. Pyrosequencing

1. Streptavidin-coated sepharose beads (Amersham Biosciences, Piscataway, NJ).
2. Binding buffer: 10 m*M* Tris-HCl, pH 7.6, 2 *M* NaCl, 1 m*M* EDTA, and 0.1% (v/v) Tween-20.
3. Denaturing solution: 0.2 *M* NaOH.
4. Washing buffer: 10 m*M* Tris-acetate, pH 7.6.
5. Annealing buffer: 20 m*M* Tris-acetate, pH 7.6, 2 m*M* $MgAc_2$.
6. Sequencing primer: 5'-TGGTACCTTTCTATTCTCAC-3' (Sigma-Genosys) (*see* **Note 2**).

3. Methods

To search for a potential ligand–receptor interaction on a particular cell type, a monolayer of the cells must be grown. Phage from a library are then added and incubated with the cells so that peptides displayed by the phage can interact and bind with any receptors on the cell surface. Internalization of phage into the cell via the receptor is minimized by incubating at 4°C. The unbound nonspecific phage are washed away and the bound phage are eluted and amplified. These amplified phage are titered and then reapplied to a monolayer of the cells using the same number of phage as before but the stringency is increased by reducing the time that the phage have to bind with receptors and also adding detergents to the wash solution. This leaves only tightly bound and specific phage. Again the phage are eluted, amplified, titered, and reapplied to cells and incubated with increasing levels of stringency. These rounds of panning continue as required (although three rounds of panning are often enough) to sufficiently enrich a population of phage from which a peptide or peptides of interest can be identified.

After the biopanning procedure and amplification of the remaining phage, the peptides displayed need to be analyzed by sequencing the randomized 21-bp region of the phage genome encoding the 7 amino acid peptide (*see* **Note 3**). Pyrosequencing can be used to sequence this region in 96 individual phage within 45 min and at the fraction of the cost associated with conventional Sanger sequencing. Because the sequences before and after the 21-bp of interest are known, primers (one of which is labeled with biotin) can be designed to produce a sequencing template using PCR. A single-stranded sequencing template can be isolated by denaturing the PCR products and rescuing the biotinylated strand. A sequencing primer is then designed and annealed to the template. The template–primer is then placed into the Pyrosequencing instrument in a 96-well plate format.

3.1. In Vitro Biopanning on Cells

1. The cells of interest are seeded in six-well plates in the relevant medium and serum and left to grow until a confluent monolayer is formed in an incubator set at the appropriate conditions for the cell line. This protocol assumes that a cell line is being used that can attach to the surface of a well and form a monolayer (*see* **Note 4**).
2. The medium is replaced with serum-free medium for 2 h.
3. The serum-free medium is then removed and replaced with 1% BSA for 1 h.
4. The BSA is aspirated away and the cells are washed gently with TBS twice. Ensure to pipet down the side of the well to reduce shear forces and damage to the monolayer.
5. 2×10^{11} phage from the PhD-7 phage display peptide library are mixed in 1 mL TBS and added to the cell monolayer in the well of the plate.
6. The plate is wrapped in Parafilm and incubated at 4°C for 60 min.
7. Nonspecific phage are removed by aspirating away the phage and washing the monolayer three times using TBS.
8. Bound phage are rescued by adding 1 mL of 0.2 *M* glycine-HCl (pH 2.2) to the cells for 10 min at room temperature.
9. The 1 mL of 0.2 *M* glycine-HCl containing the rescued phage is then transferred to a microcentrifuge tube and neutralized by adding 150 μL of 1 *M* Tris-HCl (pH 9.1) and mixing gently by inverting the tube three times.
10. The rescued phage are then amplified and titered (as described in **Subheadings 3.2.** and **3.3.**, respectively) for subsequent rounds of panning.
11. Usually at least three rounds of panning are carried out to ensure sufficient enrichment of the phage population. Each round is carried out with increasing stringency. A second round is carried out by adding 2×10^{11} phage from the now amplified and titered phage that were collected from the first round of biopanning. The process is repeated as before on a new monolayer of the same cell type but the incubation of the phage with the cells is reduced from 1 h to 15 min. In addition, 0.1% Tween-20 detergent is added to the TBS used for washing away the unbound nonspecific phage.
12. The phage rescued from the second round of panning are again amplified and titered.
13. In the third round of biopanning, 2×10^{11} phage are again applied to a fresh monolayer and the process is repeated as described previously except that the stringency is again increased by further reducing the incubation time of phage with cells to 10 min and adding 0.2% Tween-20 detergent to the TBS used for washing away unbound or weakly bound phage.
14. The rescued phage are amplified and titered ready for DNA analysis or for use in subsequent rounds of panning if required.

3.2. Amplification of Rescued Phage After Biopanning

1. ER2738 *E. coli* is streaked on a selective agar plate containing 40 μg/mL of tetracycline and incubated overnight at 37°C.

2. A single colony is picked from the plate and used to inoculate a 5-mL starter culture containing 20 µg/mL tetracycline. The starter culture is placed in a shaking incubator overnight at 37°C and rotating at 220 rpm.
3. The starter culture is diluted 1:100 in 20 mL of LB in a 250-mL Erlenmeyer flask.
4. The eluate containing the rescued phage from a round of biopanning is added to the culture and incubated at 37°C with vigorous shaking for 4.5 h.
5. The culture is transferred to a centrifuge tube and spun at 12,000*g* for 10 min.
6. The supernatant is carefully transferred to a fresh centrifuge tube and spun at 12,000*g* for 10 min.
7. Using a pipet, the upper 80% of the supernatant is collected and placed in a fresh tube and one-sixth the volume of PEG/NaCl solution is added and mixed by inverting the tube at least six times.
8. Leave the solution at 4°C overnight for phage to precipitate.
9. The solution is spun at 12,000*g* for 15 min at 4°C. The supernatant is poured away as gently as possible in order to avoid disturbing the pellet.
10. The tube is then briefly respun and any residual supernatant is carefully removed using a pipet.
11. The pellet is resuspended in 1 mL of TBS and transferred to a microcentrifuge tube for centrifugation for 5 min at 4°C.
12. The phage-containing supernatant is transferred to a fresh tube being careful not to carry over any residual cells that may have formed a pellet.
13. One-sixth volume of PEG/NaCl is added to the collected supernatant and mixed thoroughly but gently. The tube is incubated on ice for 1 h.
14. The tube is microcentrifuged for 10 min at 4°C. The supernatant is carefully discarded avoiding disruption of the pellet and briefly respun. Any residual supernatant is removed using a pipet.
15. The phage pellet is resuspended in 200 µL of TBS containing 0.02% $NaNa_3$.
16. The tube is spun in a microcentrifuge for 1 min and the supernatant containing the amplified and purified phage is transferred to a fresh tube.

3.3. Phage Titering

1. Inoculate 5 mL LB with a single colony of ER2738 *E. coli*. Incubate the culture at 37°C with shaking until the bacteria have reached a mid-log phase. This is monitored using a spectrophotometer to measure the optical density at a wavelength of 600 nm. When the optical density has reached 0.5 then the bacterial culture is in the mid-log phase (*see* **Note 5**).
2. While the bacterial culture is growing, melt some agarose top in a microwave and make 3-mL aliquots in 30-mL tubes. A range of dilutions of the phage will be used to measure the titer and three tubes of agarose top are needed for each expected dilution. Tenfold dilutions covering a range between 10^8 and 10^{11} are usually sufficient but it is possible that greater ranges are required, so make up a sufficient number of tubes containing agarose top.
3. Place the tubes containing the melted agarose top in a water bath set at 45°C ready for use.

4. Also, while the bacterial culture is growing, preheat LB/IPTG/X-gal plates in an incubator at 37°C. Again, three plates per expected dilution are required.
5. 10-fold serial dilutions are carried out of the phage in LB. Make sure to include the ranges of 10^8 to 10^{11}. It may be the case that the amplified phage are of very high or low titer in which case a range of 10^8 to 10^{11} may not be sufficient. Thus, dilute accordingly above or below this range.
6. Once the bacterial culture has reached the mid-log phase, aliquot 200 μL into microcentrifuge tubes, three per dilution.
7. 10 μL of each dilution is added to each tube containing the 200 μL of bacteria. The tubes are then capped and vortexed briefly. The tubes are now left to incubate for 5 min at room temperature.
8. Using a pipet, withdraw the infected bacteria from the microcentrifuge tube and mix with the preheated agarose top. Immediately vortex the agarose top solution briefly but thoroughly and pour onto a preheated LB/IPTG/X-gal plate. Be sure to tilt the plates so that the agarose top is evenly distributed.
9. Once the agarose top has set, the plates are incubated at 37°C overnight.
10. After the incubation period, the titer of the phage is calculated by counting the resultant blue colonies on each plate and multiplying this number by the dilution factor used and then by 100 to give the number of phage per milliliter. Because three plates have been prepared for each dilution, an average titer can be calculated for greater accuracy.

3.4. Amplification of Phage From an Infected Colony

Having completed the biopanning process, individual phage from the remaining amplified population now require to be analyzed to determine the amino acid sequence being expressed on their coat. To achieve this, a bacterial colony infected with a phage requires culturing. The source of this infected colony can be a plate from the final titration of phage after the last round of panning and amplification of remaining phage.

1. An agar plate containing 20 μg/mL tetracycline is streaked with ER2738 *E. coli* and left to incubate at 37°C overnight.
2. A colony is picked from the plate and used to inoculate 5 mL of LB containing 40 μg/mL tetracycline. The culture is placed in a shaking incubator at 37°C overnight.
3. The overnight culture is diluted 1:100 and 1 mL is dispensed into 20-mL sterilin tubes.
4. A plate containing well-spaced blue colonies of phage-infected *E. coli* from the final round of titering of amplified phage is selected. Alternatively, if phage from a fresh plate of infected bacteria are preferred, then *E. coli* can be infected as described previously and plated on LB/IPTG/X-gal plates at a suitable dilution.
5. A blue plaque is stabbed using a sterile wooden pick and this is then placed in one of the 20-mL sterilin tubes containing the diluted overnight culture of *E. coli.* This process is repeated depending on the number of phage that are required to be analyzed.

6. The tubes are placed in a shaking incubator for 4.5 h at 37°C.
7. The cultures are transferred to a microcentrifuge tube and spun for 30 s.
8. The supernatant is decanted into a fresh microcentrifuge tube and is spun for another 30 s.
9. The upper 80% of the supernatant is extracted using a pipet and transferred to a fresh microcentrifuge tube. This represents an amplified phage stock that can be stored at 4°C for up to 7 wk or can be mixed with glycerol in a 1:1 ratio and stored at –20°C for long-term storage.

3.5. Generation of Sequencing Template by PCR

A PCR approach is used to amplify the 21-bp region of interest in the phage genome that encodes the 7 amino acid peptide displayed on the surface of the phage. One primer is biotinylated (reverse primer) so that a single-stranded sequencing template can be isolated later on. The primers are designed so that they flank the 21-bp sequence of interest but also include a region 5' to the randomized region to allow for annealing of a sequencing primer. Using the primers described here, a PCR product of 107 bp is generated.

1. The PCR reaction is carried out in 0.5-mL autoclaved PCR tubes. Each tube contains the following mix in a total of 50 μL: 1X PCR Gold Buffer, 2 m*M* $MgCl_2$, 0.2 m*M* dNTP mix, 0.2 μ*M* forward primer, 0.2 μ*M* reverse primer, 1.25 U Amplitaq Gold, and 1 μL of an amplified preparation of an individual phage, as described previously.
2. The PCR reaction is carried out using a thermocycler with a heated lid. An initial step is programed where the phage are lysed at 95°C for 5 min.
3. Immediately after this, 45 cycles of 94°C for 30 s, 58°C for 30 s, and 72°C for 60 s are completed.
4. A final step of 72°C for 10 min is carried out.
5. The products of the PCR reaction require to be analyzed to ensure that the correct product is obtained and no other bands are present because of contamination. Ten microliters of the PCR mix can be loaded on to a 3% gel for analysis by gel electrophoresis. A 107-bp product should be observed with no other bands visible (**Fig. 1**). It may not be necessary to examine all the PCR reactions; usually a few will give an idea of how successful the reaction has been.

3.6. Purification and Preparation of Single-Stranded Sequencing Template and Pyrosequencing

Having produced a double-stranded PCR product that contains the 21-bp sequence of interest and a region 5' to this where a sequencing primer can anneal, a single-stranded template is required. To do this, the strong biotin–streptavidin interaction is exploited via the biotin label on the sequencing template strand. The sequencing template is then ready to be annealed with the sequencing primer and for the Pyrosequencing to begin. The following proto-

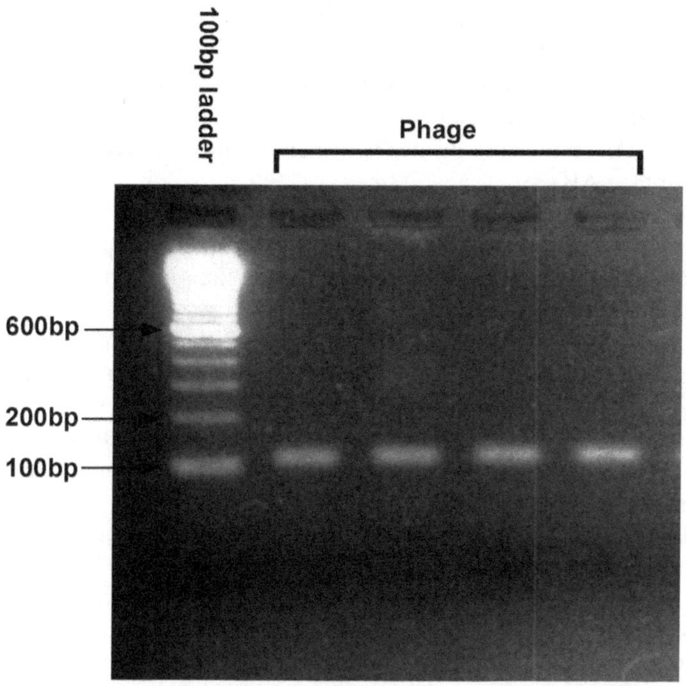

Fig. 1. PCR amplification of the randomized region of four isolated phage. PCR is carried out to amplify the randomized region of isolated phage using two primers, one of which is labeled with biotin. A product of the expected size (107 bp) is observed using agarose gel electrophoresis. The product also includes an upstream region to which a sequencing primer can anneal.

col assumes that 96 phage are being sequenced in a 96-well plate. Lower numbers will require volumes to be adjusted as required.

1. 40 μL of each PCR reaction is dispensed into a 96-well plate.
2. The streptavidin-coated beads are vortexed thoroughly to ensure that they are evenly suspended in solution. 384 μL of the beads are mixed with 3840 μL of binding buffer (*see* **Note 6**).
3. 40 μL of the streptavidin-binding buffer solution is added to each well of the 96-well plate (*see* **Note 7**) containing the PCR products and incubated at room temperature for 10-min shaking at 12,000 rpm.
4. The samples are transferred to a 96-well filter plate (Millipore) and placed on a vacuum manifold (Biotage AB) where vacuum is applied until all the liquid passes through the membrane.
5. 50 μL of denaturing solution is added to the remaining beads in the wells of the plate and left to incubate for 1 min. The vacuum is applied again until the liquid is drawn through the membrane.

6. The beads are washed twice by adding 150 μL of washing buffer while applying the vacuum.
7. 50 μL of annealing buffer is added to the wells of the plate and the beads are resuspended by using a pipet.
8. 5 μL of 3 μ*M* sequencing primer is dispensed into the wells of a PSQ 96 Plate Low (Biotage AB).
9. The 50 μL of resuspended beads is transferred from the 96-well filter plate to the wells of the PSQ 96 Plate Low containing the sequencing primer. Mixing of the beads and the primer is gently facilitated using the pipet while being careful not to contaminate neighboring wells.
10. The plate is heated to 80°C for 2 min using a PSQ Sample Prep Thermoplate Low (Biotage AB) and then left at room temperature for 10 min.
11. A dispensation cartridge (Biotage AB) for a PSQ 96MA System (Biotage AB) is prepared and filled with the required enzymes, substrates, and nucleotides from a PSQ SNP 96 reagent kit (Biotage AB) according to the manufacturer's protocol and inserted into the sequencer.
12. The plate containing the samples is placed inside the sequencer. The order of dispensation of nucleotides is entered into the software as TCT followed by cyclic rounds of all four nucleotides (*see* **Note 8**). TCT is entered first because these nucleotides are known to be present just before the randomized region begins and incorporate into the sequencing template strand as part of the elongation process. The sequencing primer was specifically designed to begin the sequencing process at this site.
13. The software is set to analyze all 96-wells of the plate and the progress is monitored in real time with the option to pause or stop the reaction at any point.
14. When the sequencing reaction is complete, the results in the form of a Pyrogram can be analyzed on the screen or printed out onto paper (**Fig. 2**). Apart from manually reading the sequence from the Pyrogram, an automated sequence is also provided.
15. All 96 sequences obtained are now checked to look for any that reoccur. The first three nucleotides should be TCT and then the next 21 bp should be the sequence of the randomized region in the phage genome. After the random 21 bp, the next three nucleotides should be GGT (as shown in **Fig. 2**). The TCT and the GGT before and after the 21-bp randomized region, respectively, act as a guide to whether the sequencing is accurate and has worked and also that the sequence is in frame and that no additions or deletions have occurred within the randomized region.

4. Notes

1. Many different types of phage display libraries are available. New England Biolabs sells other phage libraries in which the phage express a constrained 7-mer peptide or a 12-mer peptide. The PCR and sequencing oligonucleotides used in this protocol can be used for these libraries because only the randomized region changes in sequence or length, the flanking regions are the same. However, the ability to sequence over longer lengths using Pyrosequencing will require investigation.

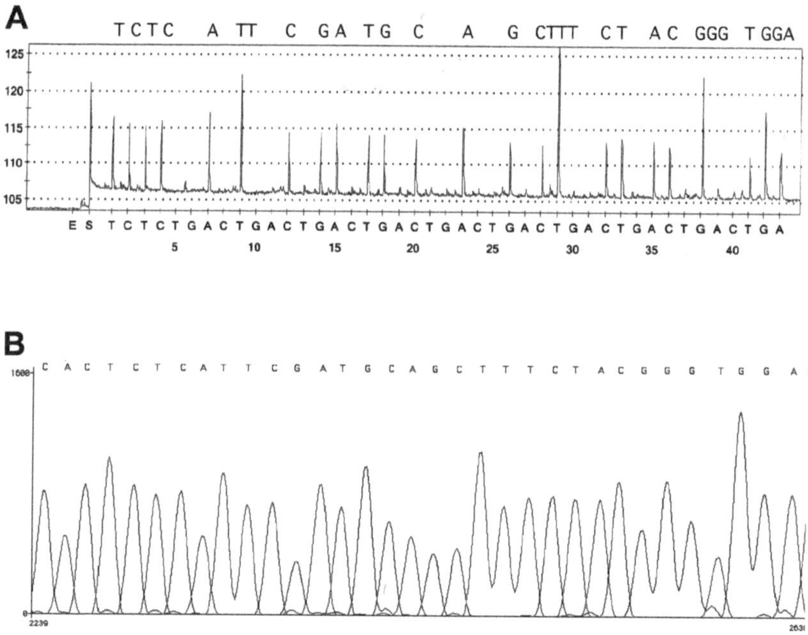

Pyrogram Sequence: **TCT**CATTCGATGCAGCTTTCTACG**GGT**

Automated Sequence: **TCT**CATTCGATGCAGCTTTCTACG**GGT**

C Peptide Sequence: H S M Q L S T

Fig. 2. Pyrosequencing® analysis of an isolated phage. **(A)** A Pyrogram® of a phage showing the sequence of the randomized region encoding the peptide displayed on the phage coat and flanking sequences. In addition to the automated predicted sequence generated by the software, the Pyrogram can be read manually as shown above the peaks. **(B)** The accuracy of Pyrosequencing for the analysis of phage is demonstrated by sequencing the same phage using traditional automated Sanger sequencing. Both methods produce exactly the same sequence. **(C)** Translation of the identical DNA sequences produced by both methods reveals that this particular phage expresses the peptide sequence histadine (H), serine (S), methionine (M), glutamine (Q), leucine (L), serine (S), and threonine (T) on its coat.

2. If using a completely different library, then the PCR primers and sequencing primer will need to be designed accordingly.
3. The PCR product can be longer than is actually required. In this case, a 107-bp PCR product is generated. This includes more sequence than just the 21-bp sequence of interest and the region for the sequencing primer to anneal. By generating a longer PCR product it is easier to visualize by agarose gel electrophoresis using only a little of the PCR mix.

4. The biopanning process can also be modified and carried out in an in vivo environment as described by Pasqualini et al. *(7)* and Arap et al. *(8)*. The analysis of the final phage population can be carried out using Pyrosequencing.
5. When titering phage, it is advisable to check the plates for signs of white bacterial colonies. Such colonies may be infected with environmental phage and so represent contamination of the phage library.
6. When making up the streptavidin-coated beads-binding buffer solution, it is best to make up a little more than is exactly required. Because of the large numbers of phage to be analyzed and small pipeting errors, more of the solution may be used up than expected.
7. It is advisable to use a multichannel pipet when using 96-well plates. This reduces labor and also reduces the chances of error.
8. Long homopolymer stretches of nucleotides have been observed to cause problems in Pyrosequencing *(6)*. In such a case it may be necessary to sequence such phage using conventional automated Sanger sequencing.

Acknowledgments

The author would like to thank Professor Charles Coutelle and Dr. Richard Harbottle at Imperial College London for their technical advice, and Yvonne Kienast for her encouragement and suggestions.

References

1. Smith, G. P. (1985) Filamentous fusion phage: novel expression vectors that display cloned antigens on the virion surface. *Science* **228,** 1315–1317.
2. Fairbrother, W. J., Christinger, H. W., Cochran, A. G., et al. (1998) Novel peptides selected to bind vascular endothelial growth factor target the receptor-binding site. *Biochemistry* **37,** 17,754–17,764.
3. Healy, J. M., Murayama, O., Maeda, T., Yoshino, K., Sekiguchi, K., and Kikuchi, M. (1996) Peptide ligands for integrin alpha v beta 3 selected from random phage display libraries. *Biochemistry* **34,** 3948–3955.
4. Jost, P. J., Harbottle, R. P., Knight, A., Miller, A. D., Coutelle, C., and Schneider, H. (2001) A novel peptide, THALWHT, for the targeting of human airway epithelia. *FEBS Lett.* **489,** 263–269.
5. Nicklin, S. A., White, S. J., Watkins, S. J., Hawkins, R. E., and Baker, A. H. (2000) Selective targeting of gene transfer to vascular endothelial cells by use of peptides isolated by phage display. *Circulation* **102,** 231–237.
6. Rahim, A., Coutelle,, C. and Harbottle, R. (2003) High-throughput Pyrosequencing of a phage display library for the identification of enriched target-specific peptides. *Biotechniques* **35,** 317–324.
7. Pasqualini, R. and Ruoslahti, E. (1996) Organ targeting in vivo using phage display peptide libraries. *Nature* **380,** 364–366.
8. Arap, W., Kolonin, M. G., Trepel, M., et al. (2002) Steps toward mapping the human vasculature by phage display. *Nat. Med.* **8,** 121–127.

12

Gene Copy Number Detection in Animal Studies

Gerli Pielberg and Leif Andersson

Summary

Sensitive methods for the quantification of DNA fragments can be used to study an individual's genetic constitution for duplicated regions of the genome or to determine the relative proportion of different DNA fragments in heterogeneous samples such as pooled DNA from different individuals or in samples in which a fraction of the chromosomes carry a mutation. Here, we describe how we are using Pyrosequencing® for this purpose. In **Subheading 3.**, we describe a sensitive method that can be used to quantify the relative proportion of X- and Y-carrying sperm after sperm sorting in cattle. We also discuss our method for determining the copy numbers at a duplicated locus. This method has been applied to study genetic variation at the *KIT* locus in pigs, which have a major effect on coat color variation in this species.

Key Words: Duplication; quantification; sperm sorting; cattle; pig; Pyrosequencing®.

1. Introduction

Large segmental duplications constitute up to 5% of mammalian genomes *(1,2)*. These duplications can span large genomic distances and may contain both high-copy number repeats and gene sequences with exon–intron structures. The sequence identity between duplicated copies can be very high, creating a substrate for recombination events and large-scale chromosomal rearrangements, such as deletions, inversions, translocations, and additional duplications. These rearrangements have an impact on the evolution of genome architecture, the emergence of new genes, and may therefore affect adaptive evolution. Segmental duplications can also be a major complicating factor in the molecular characterization of a chromosomal region, for instance during positional cloning experiments.

Segmental duplications in the human are associated with chromosomal rearrangements causing genetic diseases. In fact, it has been estimated that 1 in

From: *Methods in Molecular Biology, vol. 373: Pyrosequencing® Protocols*
Edited by: S. Marsh © Humana Press Inc., Totowa, NJ

every 1000 human births has a duplication-mediated germline rearrangement *(1)*. For instance, human cytochromes and chemokines, enzymes involved in metabolism of endogenous and exogenous compounds, are known to have interindividual and interethnic variabilities in copy numbers. This leads to phenotypical differences in the response and toxicity of many drugs (e.g., antidepressants, neuroleptics, and cardiovascular agents) and environmental compounds (e.g., nicotine) and may influence disease susceptibility *(3–5)*. Therefore, fast and easy-to-use methods for analyzing copy number variations are of considerable interest.

Segmental duplications have not yet been extensively studied in domestic animals but we have shown that one of the major coat color variants in the domestic pig, the dominant white color, is associated with both a 450-kb duplication containing the entire coding sequence of the KIT receptor and a splice mutation leading to exon skipping *(6–8)*. We have used Pyrosequencing® to quantify copy number (one, two, or three per haplotype) and to estimate the number of copies (one or two per haplotype) carrying the splice mutation *(9,10)*. In this chapter, we discuss these previously described methods of allele and gene copy number quantification, and we present a case study for quantification of sperm cells carrying X or Y chromosomes. Furthermore, in the extensive notes section, we share our experiences of using Pyrosequencing for quantification experiments.

2. Materials

2.1. DNA Samples

Samples used in this study were either pure genomic DNA prepared from blood samples by standard methods or crude lysates prepared from semen samples using Pronase E (Merck KGaA, Damstadt, Germany). Crude lysates were also further purified using QIAamp® DNA Blood Mini Kit (Qiagen, Crawley, UK) (*see* **Note 1**).

2.2. PCR Analysis

PCR reactions were carried out in a total volume of 25 μL containing 40 ng genomic DNA, 1.5 m*M* $MgCl_2$, 50 m*M* KCl, 10 m*M* Tris-HCl (pH 8.3), 200 μ*M* dNTPs, 0.7 U AmpliTaq Gold DNA polymerase (PE Applied Biosystems, Foster City, CA) and 5 pmol of both forward and reverse primer (*see* **Notes 2** and **3**).

2.3. Pyrosequencing

Pyrosequencing was carried out by immobilizing the biotinylated PCR product (25 μL) onto streptavidin-coated sepharose beads (Amersham Biosciences, Uppsala, Sweden) using binding buffer (5 m*M* Tris-HCl, 1 *M* NaCl, 0.5 m*M*

EDTA, 0.5% Tween-20, pH 7.6) and shaken for 10 min at room temperature. Single-stranded PCR template was obtained by sucking the beads up to the vacuum filters, followed by washing with 70% ethanol, denaturation with 0.2 *M* NaOH, and a final washing with 10 m*M* Tris-acetate (pH 7.6). Beads were released into 12 μL annealing buffer (20 m*M* Tris-acetate, 2 m*M* $MgAc_2$, pH 7.6) containing sequencing primer (0.3 μ*M*) and the annealing step was performed at 80°C for 2 min followed by subsequent cooling down to room temperature. Pyrosequencing was carried out using the PSQ HS 96A instrument and the Pyro Gold CDT Reagents (Biotage AB) (*see* **Notes 4–8**).

3. Methods

Methods for sorting X and Y chromosome-bearing sperms are of considerable interest in domestic animals. This is because there is often a strong preference for one sex in many production systems. For instance, a dairy farmer uses cows to produce milk and, therefore, a female calf is worth more than a male calf; on the contrary in beef production male progeny are preferred because they show a higher degree of muscularity. Thus, many methods for sperm sorting have been developed. We have therefore developed a method based on Pyrosequencing to quantify the relative proportions of X- and Y-bearing sperm in samples that have been subjected to sorting.

PCR and sequencing primers were designed using an alignment of the bovine Y-linked zinc finger protein (*ZFY*) and X-linked zinc finger protein (*ZFX*) genes (accession numbers AF032867, AF032866), in regions with 100% sequence identity between *ZFX* and *ZFY*. Sequencing primer was designed to a region that gave a dispensation order suitable for quantification. Primer sequences were the following:

ZFF: 5'-AGAAAGCAAAACACACCAGTG-3'
ZFR: 5'-Bio-CCTTTGTGTGAACTGAAATTATG-3'
ZFSeq: 5'-CATTGTGACCACAAGAGTTC-3'

PCR product size was 92 bp and PCR was performed at an annealing temperature of 55°C for 5 cycles and 52°C for 25 cycles (for additional information, *see* **Subheading 2.2.**). The dispensation order for Pyrosequencing is shown in **Table 1**. PCR products from *ZFX* and *ZFY* were cloned and these clones were mixed in different proportions in order to generate standard curves for quantification. The peak heights of two different unique peaks per template were measured and the ratio of *ZFY*/(*ZFY* + *ZFX*) was calculated and plotted against the initial proportion of cloned templates used for PCR (**Fig. 1**). The standard curves that were generated showed an excellent linearity demonstrating that we have established a sensitive method for quantifying the relative copy number of X and Y chromosomes in DNA samples from cattle.

Table 1
Dispensation Order Used for Pyrosequencing[a]

Gene	A	C_1	G	T_1	C	A	T	C_2	A	T_2	G	C	A	G	T	G	C
ZFX	—	—	G	—	—	AA	—	C	—	T	—	C	AA	G	—	—	C
ZFY	AAA	C	—	T	C	—	T	—	A	—	G	—	—	—	T	G	—

[a]C_1 and T_1 indicate the two nucleotide positions in *ZFY* used for quantification, whereas C_2 and T_2 were used for *ZFX*.

3.1. Discussion

We have demonstrated that Pyrosequencing can provide an excellent quantification of incorporated nucleotides. As we have described in **Subheading 3.**, Pyrosequencing can be used for quantification of different templates/chromosomes, for instance, for analyzing the proportions of sperm cells carrying an X or Y chromosome. In a previous study we used Pyrosequencing for genetic analysis of the highly polymorphic *KIT* locus in pigs controlling different coat color variants ***(9,10)***. Previous studies had revealed that the *KIT* gene is duplicated in white pigs ***(6)*** and that one of the copies carry a splice mutation that caused exon skipping and a KIT receptor lacking a critical region of the tyrosine kinase domain ***(7)***. First, we used Pyrosequencing to quantify the relative proportion of normal copies vs copies carrying the splice mutation ***(9)***. These studies showed that a *KIT* haplotype in pigs may carry one, two, or three copies of a duplicated segment and that one or two of these may carry the splice mutation. The cloning of the duplication breakpoint ***(8)*** allowed us to develop a sensitive Pyrosequencing method for quantifying the copy number ***(10)***. The design of the method is outlined in **Fig. 2**. We used a shared forward primer (F) and a shared sequencing primer (Seq) but different reverse primers (R1 and R2) that target the internal breakpoint and a flanking breakpoint. The F and R1 primer pair will give one amplification product per chromosome and the F and R2 primers will give one product per duplicated copy. This means that a heterozygous individual with a normal allele and a duplicated allele will give a 2:1 ratio between the F/R1 product and the F/R2 product (*see* **Fig. 2**). A 1:1 ratio will be obtained for individuals that are homozygous for a duplicated allele or heterozygous for a normal allele with one copy and an allele with three copies.

In our initial experiments we got a poor linearity using the F, R1, and R2 primers in the same reaction despite the fact that the reverse primers were used in excess in an attempt to avoid that one of these became rate limiting. We therefore modified the design and introduced tailed primers and a shared-tail primer (**Fig. 2**). The R1 and R2 primers were used in minute amounts to spark the reaction and the consequence of this design was that the amplicons gener-

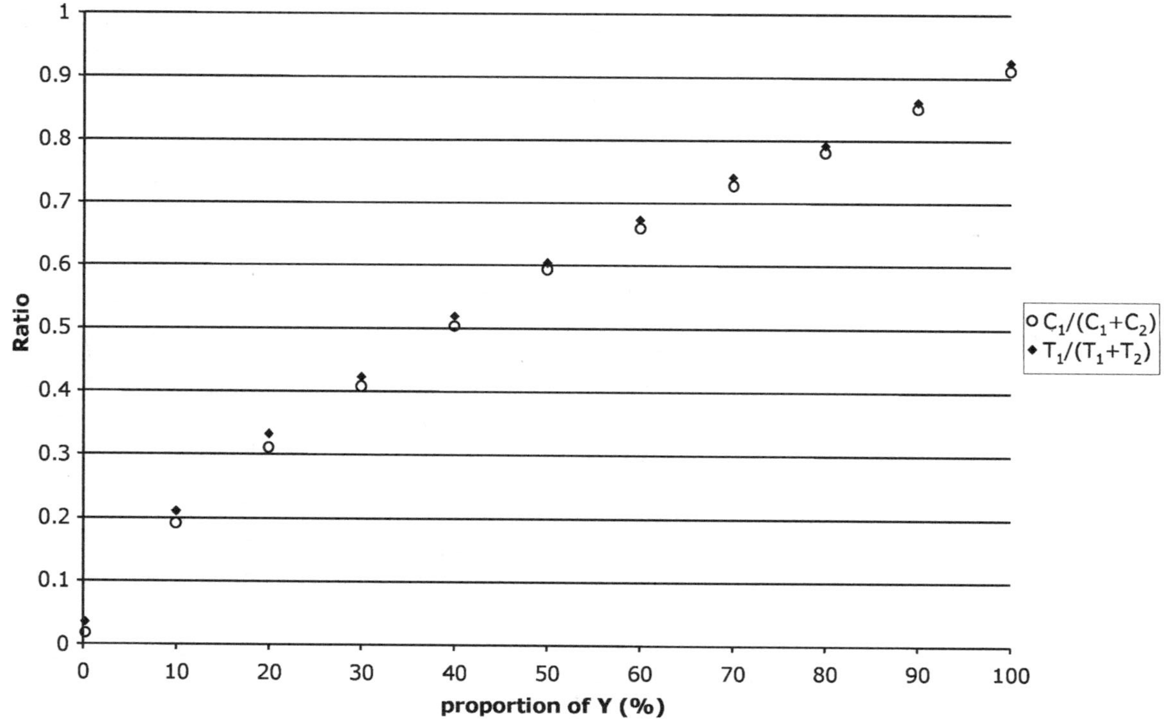

Fig. 1. Standard curves generated using cloned *ZFX* and *ZFY* PCR products. The circle and the diamond symbols indicate two different ratios calculated using unique *ZFX* and *ZFY* peak heights.

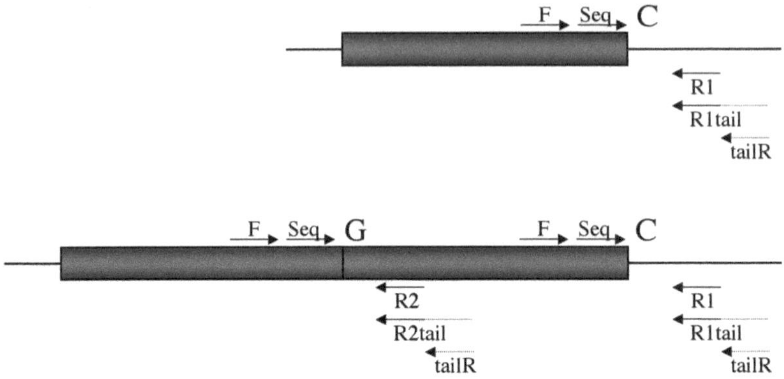

Fig. 2. Design of a test quantifying the gene copy number at the porcine *KIT* locus. The normal copy was amplified using the shared forward primer (F) and a specific reversed primer (R1), or alternatively a mix (1:100) of the two reversed primers R1 tail and tail R. The duplication breakpoint was amplified using F and a specific reversed primer (R2), or a mix (1:100) of R2 tail and tail R. Primer Seq was used for Pyrosequencing.

ated from the internal and the flanking breakpoints were competing for the same primer pair. Using this approach we obtained an excellent linearity and we were able to resolve an extensive genetic variation at this locus in pigs ***(10)***.

The downside of this method is that one needs prior information of the duplication breakpoint. Furthermore, our experiences show that the method gives a good resolution to distinguish individuals with up to five copies, but as the copy number increases it becomes gradually more and more difficult to accurately distinguish genotype classes that differ by only one copy. This is because the relative increase in the signal from the breakpoint becomes smaller and smaller.

4. Notes

1. The quality of the template is very important for precise quantification. Crude lysates and other samples containing possible PCR inhibitors may not give a reproducible quantification result and should therefore be purified. An insufficient amount of template for PCR may also result in a skewed quantification ratio. The optimal way to evaluate the test and the quality of samples would be to use different preparations of the same template as well as separate PCR runs of the same template for comparison.
2. Primer design is crucial for a good Pyrosequencing assay. PCR primers should be designed as normal PCR primers. It should be considered that the longer the amplified fragment, the more chances there are for all possible secondary structures that might cause problems in analysis. PCR primers should not be designed in repetitive sequences and other sequences that might provide an alternative target sequence (for instance pseudogenes). If different quantifiable targets are

amplified by different primers, their melting temperature should be as similar as possible and the amplicons that they amplify should have as similar a length and nucleotide content as possible, in order to minimize preferential amplification of one of the templates. If a combination of tailed primers and tail-specific primer is used, tail sequence should not align anywhere on the genome of the species of interest. Mutations in the PCR- and sequencing-primer sites have crucial effects on quantification results. If possible, ambiguous positions in the primer sites should be avoided. Sequencing primer should be designed not too far from the quantifiable peaks (1–15 nt would be optimal) and should provide a suitable dispensation order for quantification. Furthermore, sequencing primer should not align anywhere else in the amplicons and not form a strong dimer with itself.

3. Any preferred PCR system could be used for amplification. If the amplicons are GC-rich or otherwise hard to amplify, we have used chemical denaturating agents (5% [v/v] DMSO or 1 *M* betaine at final concentration). Leftover biotinylated PCR primer in a PCR reaction can cause lower Pyrosequencing signals because this primer competes during the immobilization step with the PCR product. Therefore, the amount of biotinylated PCR primer that we use is 0.5–1 pmol/25 μL reaction.
4. An optimal dispensation order for quantification should provide more than one unique, i.e., quantifiable, peak for each amplicon. That gives an opportunity for comparison and statistical analysis for the evaluation and optimization of the assay. So-called sequencing-with-shift can be a good strategy for creating unique peaks. Long homopolymeric stretches (>4 nt) should be avoided in the dispensation order, or the peaks used for quantification should lie on the same side of the homopolymeric stretch. If possible, using the A-nucleotide peak for quantification should be avoided (it could be easily done by sequencing the other strand) or one should be ready to standardize peak heights before quantification in order to minimize run-to-run differences.
5. Considerations for sample preparation vary from the tool used for it (vacuum or magnetic tool). If a magnetic tool is used, it is used with a dispensable plastic cover and therefore run-to-run contamination is not a problem. If using a vacuum-prep tool, generally the filters are not changed between each run. Therefore, treating the filters with bleach (sodium hypochlorite solution with 0.5% active chlorine) and the following thorough washing with water are very important for avoiding run-to-run contamination. Furthermore, it should be considered that the sample preparation steps (especially first washing with 70% ethanol) should be thorough.
6. The most common problem that Pyrosequencing users face is possible secondary structures giving alternative priming sites on the template and thereby causing background and extra peaks in the Pyrogram® charts. A good routine for identifying these problematic secondary structures is to include the following control wells in the first run:
   a. Sequencing primer in annealing buffer without a PCR product. If there are peaks in this Pyrogram, these are caused by sequencing primer dimer and these peaks should definitely not be used for the quantification.

b. Single-stranded PCR (ss-PCR) product in annealing buffer without sequencing primer (should be done for both quantifiable templates). The peaks that appear on this Pyrogram should definitely not be used for the quantification. The reason for possible peaks appearing in this well is a priming site that the unbiotinylated end of the ss-PCR product can provide by looping back to itself and forming a secondary structure. These secondary structures can be predicted by primer design software or experienced Pyrosequencing users may spot the problem already at the primer design step.
c. Biotinylated PCR primer in annealing buffer with sequencing primer. Peaks that occur in the Pyrogram can cause a problem in samples that have not given a sufficient PCR product.

Possible solutions for users facing the abovementioned problems could be:

d. Redesign of primers. One may consider designing PCR primers that amplify a shorter fragment, designing a sequencing primer with higher melting temperature, or even add an unspecific stretch of nucleotides to the 5'-end of the unbiotinylated PCR primer. The latter works as a hanging tail of the potential priming secondary structure and thereby prevents DNA-strand synthesis.
e. Use single-strand-binding protein. It should be added after the annealing step at a concentration of 0.5–2.2 μg/well.

7. The first thing to check during the analysis of a Pyrogram is to look at the peak heights and widths. If the peaks are very wide, less PCR product should be used; if the peaks are very low, more PCR product should be used and/or the PCR reaction should be optimized. Quantification should be done using only the peaks without any background. It is best to be able to compare ratios of several different unique peaks.
8. Control samples and standard curves are a very good way for evaluating and optimizing the assay. Standard curve could be run using bacterial artificial chromosome clones or plasmids harboring different templates (if the tailed-primer approach is used, these templates should not include the tail sequence). This will not give a complete answer about how the assay works at the presence of the whole genome. Alternatively, two genomic samples and their proportional mixtures could be used for standard curve experiments. In order to make different PCR and Pyrosequencing runs easier to compare and possible problems easier to identify, control samples (from a standard curve or an already quantified samples) would be good to include in each run. Furthermore, other known methods suitable for quantification could be used to evaluate the method.

Acknowledgments

This work has been supported by The Foundation for Strategic Research and the Swedish Research Council for Environment, Agricultural Sciences, and Spatial Planning.

References

1. Eichler, E. E. (2001) Recent duplication, domain accretion and the dynamic mutation of the human genome. *Trends Genet.* **17,** 661–669.
2. Bailey, J. A., Gu, Z., Clark, R. A., et al. (2002) Recent segmental duplications in the human genome. *Science* **297,** 1003–1007.
3. Løvlie, R., Daly, A. K., Molven, A., Idle, J. R., and Steen, V. M. (1996) Ultrarapid metabolizers of debrisoquine: characterization and PCR-based detection of alleles with duplication of the CYP2D6 gene. *FEBS Letters* **392,** 30–34.
4. Townson, J. R., Barcellos, L. F., and Nibbs, R. J. (2002) Gene copy number regulates the production of the human chemokine CCL3-L1. *Eur. J. Immunol.* **32,** 3016–3026.
5. Xu, C., Goodz, S., Sellers, E. M., and Tyndale, R. F. (2002) CYP2A6 genetic variation and potential consequences. *Adv. Drug Delivery Rev.* **54,** 1245–1256.
6. Johansson Moller, M., Chaudhary, R., Hellmen, E., Hoyheim, B., Chowdhary, B., and Andersson, L. (1996) Pigs with the dominant white coat color phenotype carry a duplication of the *KIT* gene encoding the mast/stem cell growth factor receptor. *Mamm. Genome* **7,** 822–830.
7. Marklund, S., Kijas, J., Rodriguez-Martinez, H., et al. (1998) Molecular basis for the dominant white phenotype in the domestic pig. *Genome Res.* **8,** 826–833.
8. Giuffra, E., Törnsten, A., Marklund, S., et al. (2002) A large duplication associated with Dominant White color in pigs originated by homologous recombination between LINE elements flanking *KIT*. *Mamm. Genome* **13,** 569–577.
9. Pielberg, G., Olsson, C., Syvänen, A.-C., and Andersson, L. (2002) Unexpectedly high allelic diversity at the *KIT* locus causing dominant white color in the domestic pig. *Genetics* **160,** 305–311.
10. Pielberg, G., Day, A. E., Plastow, G. S., and Andersson, L. (2003) A sensitive method for detecting variation in copy numbers of duplicated genes. *Genome Res.* **13,** 2171–2177.

13

Detection of Allelic Imbalance in Gene Expression Using Pyrosequencing®

Hua Wang and Steven C. Elbein

Summary

Single-nucleotide polymorphisms (SNPs) are common in the human genome, with more than 11 million SNPs having frequencies greater than 1%. The challenge is to identify the minority of functional SNPs from the large number of SNPs that are expected to be silent. Whereas coding variants are unusual, and functional (nonsynonymous) coding SNPs likely rare, regulatory SNPs appear to be common. Traditional methods to identify these SNPs in vitro are time consuming and challenging. An alternative method is to examine the allele-specific expression in the cDNA from tissues expressing the genes of interest and in individuals heterozygous for a transcribed SNP. This method permits expression to be evaluated in the context of the same *trans*-acting factors and to identify genes with likely *cis*-acting regulatory variants or parent of origin (imprinting) effects. Such studies require a method to reliably quantify the expression from each allele. Pyrosequencing offers such capabilities, and given the relatively low cost and high throughput, it offers a sensitive method to determine allelic imbalance in the cDNA from tissues expressing genes of interest.

Key Words: Pyrosequencing; allele-specific expression; single-nucleotide polymorphisms; SNPs; allelic imbalance.

1. Introduction

The human genome includes more than 10 million single-nucleotide polymorphisms (SNPs) in which a nucleotide differs in at least 1% of individuals. Although the vast majority of this sequence variation likely has no effect on phenotype, the susceptibility to complex disease is thought to result from a subset of these SNPs. Sequence variants may affect phenotype by altering protein function, generally by changing protein sequence (nonsynonymous SNPs). However, coding variants are a small minority of SNPs. In contrast, most complex disease variants have mapped to noncoding SNPs. This fact suggests that

From: *Methods in Molecular Biology, vol. 373: Pyrosequencing® Protocols*
Edited by: S. Marsh © Humana Press Inc., Totowa, NJ

most variation contributing to disease susceptibility will alter expression rather than function of proteins. Such changes in gene expression may occur in two ways: by acting on genes at a distance through an expressed transcription factor or noncoding RNA (*trans*-acting factor), or by acting to regulate a gene on the same allele (*cis*-acting factor). Large surveys and mapping studies suggest that both *cis*- and *trans*-acting variants are common ***(1,2)***. Whereas *trans*-acting variants will be expected to alter total gene expression between individuals, *cis*-acting variants in heterozygous individuals will result in a difference in the expression of the two alleles regardless of other factors that may act on the expression of the gene. This allele-specific expression, or *allelic imbalance*, may result from an inherited *cis*-acting variant, or less often, the parent of origin effects (gene imprinting) that result in the silencing of allele expression from one parent. Although altered expression likely results most often from promoter or enhancer variants that alter transcription factor-binding sites, more recent data suggest that 3' untranslated region variants may alter RNA stability of one allele, or may alter binding of microRNA and cause altered stability or protein translation ***(3)***.

1.1. Inherited Regulation of Gene Expression in Humans

Unexpectedly, variants that have been mapped to complex disease susceptibility have not followed the pattern of known monogenic disease mutations in altering protein structure and function. Instead, variants accounting for susceptibility to T2DM ***(4–7)***, coronary disease ***(8–10)***, asthma ***(11,12)***, and others are more often noncoding than coding ***(13)***. In some cases (5-lipoxygenase ***[9]*** and Calpain 10; ***[14,15]***) the associated haplotypes or variants were associated with altered gene expression, whereas in other cases regulatory variants are suspected but not yet demonstrated.

Several large surveys have sought to determine whether such inherited changes in gene expression are in fact common. Most surveys used transformed lymphocytes as an available tissue for which both family material and dense SNP maps are publicly available. Three large surveys of this type have been published. Morley et al. ***(16)*** examined 3554 "expression phenotypes" that differed among cell lines from CEPH families, and found significant linkage to 984 expression phenotypes. They considered evidence for *cis* regulation if the locus mapped to within 5 Mb of the target gene. They reported evidence for both *trans*- and *cis*-acting regulatory variants, with *trans* accounting for nearly 80% of the loci mapped. Their work suggests a complex regulatory network in which many genes may be controlled by multiple clustered regulatory variants. Nonetheless, they could demonstrate *cis*-acting variants that resulted in differential allelic expression or allelic imbalance, with evidence for linkage disequilibrium between a SNP genotype and a nearby regulatory determinant of the expression phenotype. In a follow-up study, Cheung et al. showed an association of HapMap SNPs with the strongest of these linkages, again suggesting that variants associated with altered gene expression are common *(1)*. Monks et al. *(2)* expanded this approach by testing 15 CEPH families for 23,499 genes, of which 2340 were expressed in transformed lymphocytes. Of these, 31% showed heritable differences in expression and 33 genes showed significant linkage to a quantitative trait locus. Approximately 30% of these quantitative trait locus again mapped close to the gene of interest, suggesting *cis* action. Most recently Stranger et al. *(17)* conducted a similar experiment using the HapMap panel in 60 CEPH Caucasian samples to perform a genome-wide association of gene expression in 630 genes, again in transformed lymphocytes. Of the 374 expressed genes, the most abundant associations were near the gene of interest and thus supported a larger role for *cis* than *trans* regulation of transcript levels. As with other genome-wide association analyses, however, multiple correction methods require a threshold that likely misses many important linkages or associations of gene expression with genome association or genome scan data.

1.2. Allele-Specific Expression

The studies previously reviewed clearly demonstrate that genetic controls on gene expression exist, but considering the large number of genetic variants and the large number of expressed genes, the number of expression quantitative trait loci mapped by these methods to either regions or SNPs is quite small. In addition to the problems of correcting for multiple tests inherent in these genome-wide association or array-based approaches, total gene expression is likely the result of the integration of multiple genetic variants as well as environmental influences and even the effects, rather than cause, of the disease process *(18)*. Identification of these regulatory variants is expected to be critical in the search for complex disease genes. Regulatory variants acting on the same allele (*cis*-acting variants) might alter transcription through enhancer or promoter sequences, or RNA stability through the 3' or 5' untranslated regions. Many of these effects are likely modest, yet in aggregate these modest effects likely lead to disease states. Hence, sensitive methods are essential to identify individual effects in large numbers of genes. Comparison of levels of mRNA from maternal and paternal alleles from individuals heterozygous for a transcribed SNP offers such a method. In the absence of *cis*-acting controls on transcription or RNA stability, the ratio of maternal and paternal alleles should be one (equal amounts of each allele). Any significant deviation from this ratio suggests regulatory controls on RNA transcription or stability. Whereas the built-in control (the wild-type allele) provides for high sensitivity for allelic-imbalance studies, detection of single, heterozygous *cis*-acting variants on *to-*

tal gene expression, as would be measured in a microarray analysis, may be difficult with only one allele altered. This modest effect on total gene expression may account for the relatively low estimates of *cis*-acting variants in published surveys of gene expression previously reviewed ***(18)***.

Several large surveys have examined allele-specific expression or allelic imbalance and documented that *cis*-acting variation leading to altered mRNA levels is indeed common. An initial report from Yan et al. ***(15)*** examined transformed lymphocytes from 96 members of CEPH families, and observed significant differences in 6 of 13 genes studied with up to 30% of heterozygous individuals showing an imbalance. The altered expression was inherited in a Mendelian fashion for two variants, including the type 2 diabetes gene *CAPN10* and the gene *PKD2*. In a larger study using a similar approach, Pastinen et al. ***(19)*** selected 495 SNPs from 239 genes, from which they were able to study 193 SNPs from 126 genes and 3 known imprinted genes in transformed lymphocytes. They observed deviations in allelic expression exceeding the 1.5/1 threshold (60/40% allelic ratio) in 23 of 126 genes tested. Clearly, a limitation of these studies is the need for a transcribed SNP, but in proof of principle, Pastinen et al. also carried out allelic imbalance studies in unspliced RNA (heteronuclear or hnRNA) and showed results concordant with those from mRNA ***(19)***. The direction of the allelic imbalance was not consistent across cell lines, pointing to a caveat in the interpretation of these studies that is addressed next. The SNPs tested in this study and most analyses were at some distance from the transcription start site, and possibly even farther from putative enhancer sequences. Hence, in many cases the SNP used to detect allelic imbalance (marker SNP) and the causative or regulatory SNP (r-SNP) may be quite distant and in incomplete linkage disequilibrium. These authors also demonstrated transmission of allelic imbalance consistent with Mendelian inheritance for several but not all families. Monoallelic expression was observed for several variants, and was transmitted in a fashion most consistent with parent-of-origin (imprinting) effects. In a single gene (*BTN3A2*), concordance in allelic imbalance was also demonstrated between adipose and transformed lymphocytes, albeit not in the same individual. Both Yan et al. ***(15)*** and Pastinen et al. ***(19)*** found approximately one-fifth of expressed genes showed allelic imbalance.

Most of the studies described so far examined transformed lymphocytes, which have also been the primary tissue studied from our laboratory (*see* **Subheading 2.** ***[20,21]***). Two studies have taken an alternative approach to examine primary tissues. Bray et al. ***(22)*** examined 60 unrelated postmortem brain samples for 15 selected genes. Of these 15 genes, 7 showed allelic imbalance beyond the defined threshold of a 20% increase in the ratio in at least one individual. The conclusions were remarkably similar to those of Yan et al.

previously reviewed *(15)*, and suggest nearly 50% of genes might have *cis*-acting controls. Lo et al. *(23)* took a very different approach and used a chip-based hybridization method to examine 1494 SNPs in liver and kidney from 7 fetuses. The Affymetrix HuSNP chip with methods to quantify the ratio of expression could detect only a twofold difference in allelic balance, as opposed to the smaller ratios detected in other studies. Additionally, the small sample size clearly limits the sensitivity of this study. Finally, only 602 SNPs had at least one heterozygous individual, were in the transcribed sequence, and were in a gene expressed in liver or kidney. Despite these limitations, 54% (326) of genes showed differential expression in at least one individual *(23)*. Thus, even in a study with small sample size and limited ability to detect more subtle ratios, nearly half of the unselected genes show allelic imbalance. Clearly the need exists for more studies in other tissues and with sensitive methods.

1.3. Allelic Imbalance Studies: Caveats and Technical Considerations

The studies previously reviewed clearly demonstrate that some heterozygous individuals demonstrate different amounts of cDNA from the maternal and paternal alleles in a variety of tissues and using a variety of methods. With the exception of the HuSNP chip, which provides the opportunity for very high throughput but lower sensitivity, most methods have chosen single base extension to quantify the two alleles. Pastinen et al. *(19)* used fluorescence polarization to examine the ratio of rhodamine-110 to Tamra. Other authors used fluorescent dideoxy terminator-based methods *(15)* or SNaPshot (PE Applied Biosystems) with software measurement of peak heights. More novel methods have been proposed based on informatics approaches and expressed sequence tag (EST) mining *(24)*, using chromatin immunoprecipitation to quantify RNA polymerase II binding to each allele *(25)* and using a "polony" technique to PCR amplify genes in a cDNA library immobilized in polyacrylamide. The advantage of more technologically advanced techniques over simple, sequence-based methods is unclear. We *(20,21,26–28)* and others *(29)* have used Pyrosequencing to quantify alleles. A detailed discussion of Pyrosequencing methods will not be reviewed here, but is based on the detection of pyrophosphate (PP_i) released during DNA synthesis. PPi is converted to ATP by a mixture of sulphurylase and luciferase, with subsequent conversion of luciferase to oxyluciferin with emission of light. The amount of light, quantified as the peak height, is proportionate to the amount of incorporated nucleotide, making quantification possible. Quantification software developed to permit typing of pooled DNA samples *(30,31)* works well to quantify the amount of transcript from each allele. As with SNP typing for Pyrosequencing, this method is accurate, rapid, and can be used with a universal biotinylated primer (*see* **Subheading 3.**) to minimize costs.

Whatever method is used to detect allelic imbalance, some correction must be made for uneven amplification not related to uneven transcription. We and others *(22,32)* have used DNA, where the ratio by definition is 50:50, to normalize or compare ratios observed in cDNA. We have observed many examples where ratios are consistently 40:60 in the genomic DNA; thus, normalization is essential. Because this ratio is often assay specific, the same assay must be used to test DNA and cDNA samples. When this is not possible, often because the marker SNP is near an intron–exon junction, a standard curve must be developed by mixing various amounts of the two homozygous samples using the same SNP assay used to quantify allelic imbalance.

When considering normalized cDNA ratios, an observation common to all studies is that only a fraction of heterozygous individuals demonstrate allelic imbalance. This observation might result from variable expression based on other genetic or environmental factors, from experimental variation in the measures of allelic peak heights, or from variation in the reverse transcription efficiency of the alleles. We consider a more likely explanation. Allelic imbalance studies typically test a marker SNP that is within an exon or the 3' untranslated region. The marker SNP provides a means to distinguish alleles, but if *cis*-acting variants are present, they are typically at some distance from the marker SNP. In the presence of complete linkage disequilibrium (marker and regulatory SNPs are completely correlated; $r^2 = 1$), the marker and putative *cis*-acting regulatory SNPs are in phase and all individuals heterozygous for the marker SNP will be heterozygous also for the regulatory SNP. However, for many genes, particularly large genes, the distance between the marker SNP and regulatory SNP will be larger than the usual haplotype block size. In this case, the marker and regulatory SNPs are no longer correlated (**Fig. 1**). For a rare regulatory SNP, or a regulatory SNP that is in linkage equilibrium with the marker SNP, individuals may be heterozygous at the marker SNP but homozygous at the regulatory SNP (**Fig. 1**, genotypes 1–4 and 2–3), and hence will not be expected to show allelic imbalance. From these considerations, straightforward statistical tests of significant difference from genomic DNA ratios may miss true allelic imbalance, which will be present in only a few individuals. Most studies have required the observation of allelic imbalance in three to four individuals. Even among individuals in whom allelic imbalance is observed, however, the direction of the imbalance may be discordant (**Fig. 1**, genotypes one to two vs three to four) if the marker phase is different.

Given the problems of deciding whether a small number of samples deviate from the expected ratio, standard statistical methods are insensitive. For example, if only 4/40 heterozygotes show allelic imbalance, the overall difference from the genomic DNA will likely not be significant. Establishment of the 95% confidence intervals for the allelic ratio in DNA offers a practical means to address this issue.

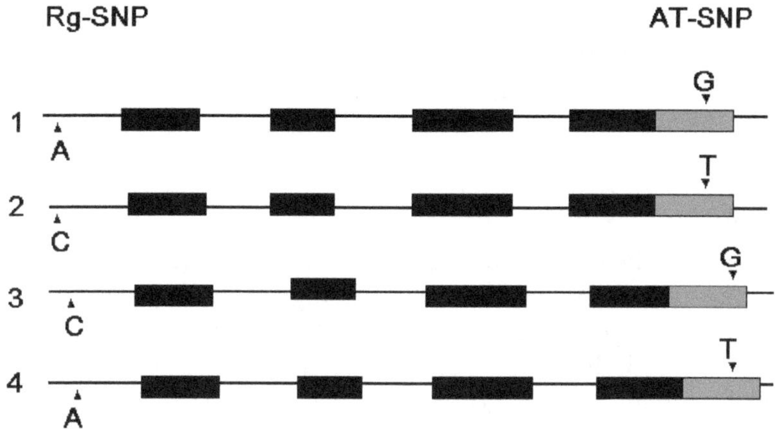

Fig. 1. Relationship between marker (allele tagging-single-nucleotide polymorphism [AT-SNP]) and regulatory SNP (Rg-SNP). The figure illustrates the text and shows the alternative phases that might be expected under incomplete linkage disequilibrium. In genotype combinations 1–2 and 3–4, both marker (AT) SNP and the Rg-SNP are heterozygous, but because the phase is different, a different allele would be overexpressed. In genotypes 2–3 and 1–4, however, the marker SNP is heterozygous but the regulatory SNP is homozygous. In this scenario, no allelic imbalance will be observed. The frequency of allelic imbalance in the absence of complete linkage disequilibrium will depend on the regulatory SNP frequency and the strength of the linkage disequilibrium.

1.4. Allelic Imbalance and Haplotypes

Identification of a marker SNP that shows different levels of transcription from the two alleles in a heterozygous individual provides only the first step in the identification of a putative *cis*-acting regulatory variant. In most cases, this marker SNP exists on an extended haplotype of correlated SNPs that define an extended region of linkage disequilibrium, or a haplotype block *(33,34)*. Within this block, often several SNPs will correlate with both total levels of gene expression in homozygous and heterozygous individuals, and in the case where multiple transcribed SNPs are present, with levels of allelic imbalance *(32)*. These "regulatory haplotypes" *(32)* in turn likely encompass one or perhaps several variants that alter either levels of transcription or allele-specific RNA stability. Identification of the actual regulatory variant will likely require in vitro experiments to test each SNP one at a time in an expression vector. However, when the full family of potential regulatory SNPs is uncovered, several informatics approaches may narrow the candidate variants. First, individuals showing allelic imbalance must be heterozygous at the unseen regulatory SNP. Hence, in individuals heterozygous for the marker SNP, any putative regulatory SNP must also be heterozygous. Secondly, regulatory SNPs may be expected to reside in regions that show conservation among species *(35)*. Third, regulatory SNPs may alter transcription factor-binding sites. These strategies can narrow the potential SNPs to be examined using time consuming and labor intensive in vitro approaches.

1.5. Sample Size Considerations

As we previously noted, individuals heterozygous for the marker SNP must also be heterozygous for the inferred regulatory variant in order for allelic imbalance to be observed. The sample size will depend on the number of individuals showing allelic imbalance that an investigator requires to move to subsequent validation steps. We propose a reasonable compromise of three individuals. We consider two extremes to illustrate the required sample size to achieve this goal. First, we consider complete linkage disequilibrium between the marker and regulatory SNP, such that the marker SNP is a complete proxy for the regulatory SNP. Hence, every individual heterozygous at the marker SNP will also be heterozygous at the regulatory SNP. We assume that estimates of allelic ratios and reverse transcription are accurate when a threshold of 1.5 is considered significant. Even with a relatively rare marker SNP at the minor allele frequency 0.1, the lower 95% confidence interval with 96 samples will be 10 heterozygous individuals, and all will show allelic imbalance. Clearly, even smaller numbers of samples ($n = 30$) in this best-case scenario would be adequate. At the other extreme, we consider complete equilibrium. The minor allele frequency of the regulatory SNP is no longer fixed by the frequency of the marker SNP. Sample size will be determined by the known minor allele frequency of the marker SNP (consider 0.1 to 0.5) and the unknown frequency of the regulatory SNP. For a marker SNP with minor allele frequency 0.2, the lower limit of detection in 96 samples for the regulatory SNP will be a minor allele frequency of 0.25. By increasing the marker SNP frequency to 0.3, a regulatory SNP with a smaller minor allele frequency of 0.15 can be detected. In the presence of complete equilibrium, practical sample sizes will limit the detection of rare regulatory SNPs. However, complete equilibrium will probably be the exception, and the ability to detect less common regulatory SNPs will in fact be much higher. Nonetheless, the analysis points out the importance of choosing marker SNPs that have high frequencies.

2. Materials

The studies we describe may be conducted on cDNA made from any tissue. Studies in our laboratory and most published studies have used Epstein–Barr virus-transformed lymphocytes. Both tissue-specific expression, and potentially tissue-specific allelic imbalance might be expected. The study of multiple tissues, or when possible tissues directly related to the disease, would be recommended. In all cases RNA is harvested and reverse transcribed. Detailed methods followed in our laboratory are described.

1. Cell lines: Epstein–Barr virus-transformed lymphocytes from the human subjects were grown to 0.5–1.0 × 10^6 cells/mL in RPMI-1640 media (Omega Scientific Inc, Tarzana, CA) with 10% fetal bovine serum and 11 m*M* glucose.
2. Tissues: although any tissue may be used, we have focused on abdominal subcutaneous adipose, which was obtained by either incisional biopsy or needle biopsy after an overnight fast and under local anesthesia. Adipose was cut into 100-mg fragments, rinsed in saline, and quick frozen in liquid nitrogen for future study.
3. 2X Binding buffer: 10 m*M* Tris-HCl, pH 7.6, 2 *M* NaCl, 1 m*M* EDTA, and 0.1% Tween-20.
4. Annealing buffer: 20 m*M* Tris-HCl, pH 7.6 and 2 m*M* MgAc.

3. Methods

3.1. Primer and Assay Design

Primers are designed using the PSQ assay design software (Biotage, AB). Consideration must be given to the predicted peaks and dispensation order. First, as noted next, the best assays will be designed so that cDNA and genomic DNA may be assessed in the same assay. SNPs very close to the intron/exon boundary cannot be assessed with the same cDNA assay because of splicing. Second, to compare ratios the result must provide only two peaks. As shown in **Fig. 2A,B**, dispensation order may need to be altered such that only two peaks vary, not three. Finally, homopolymers (more than one base in a row) give peak heights over one. Allele quantification software will still estimate the percentage of each allele (C and T in this case), but the accuracy of that estimate is likely lower when the peaks are not expected to be equal (*see* **Fig. 1**).

3.2. DNA Extraction, Genotyping, and Allele Quantification

DNA was extracted from peripheral leukocytes obtained from whole blood using the Puregene kit (Gentra Systems, Inc., Minneapolis, MN). Quality of DNA was determined by looking for high molecular weight on 0.5% agarose and staining with ethidium bromide. Quantitation of DNA was measured spectrophotometrically using the 260/280 n*M* ratio, and DNA concentration calculated from the 260 n*M* measure. DNA genotyping was performed by Pyrosequencing using a PSQ 96 system (Biotage AB, Uppsala, Sweden). Genotyping results were used to validate the quality of the SNP, determine the allele frequency in unrelated individuals, and to identify the heterozygous individuals for allelic imbalance studies. Additionally, heterozygous individuals provided the samples to control for allelic imbalance resulting from unequal amplification or other assay artifacts.

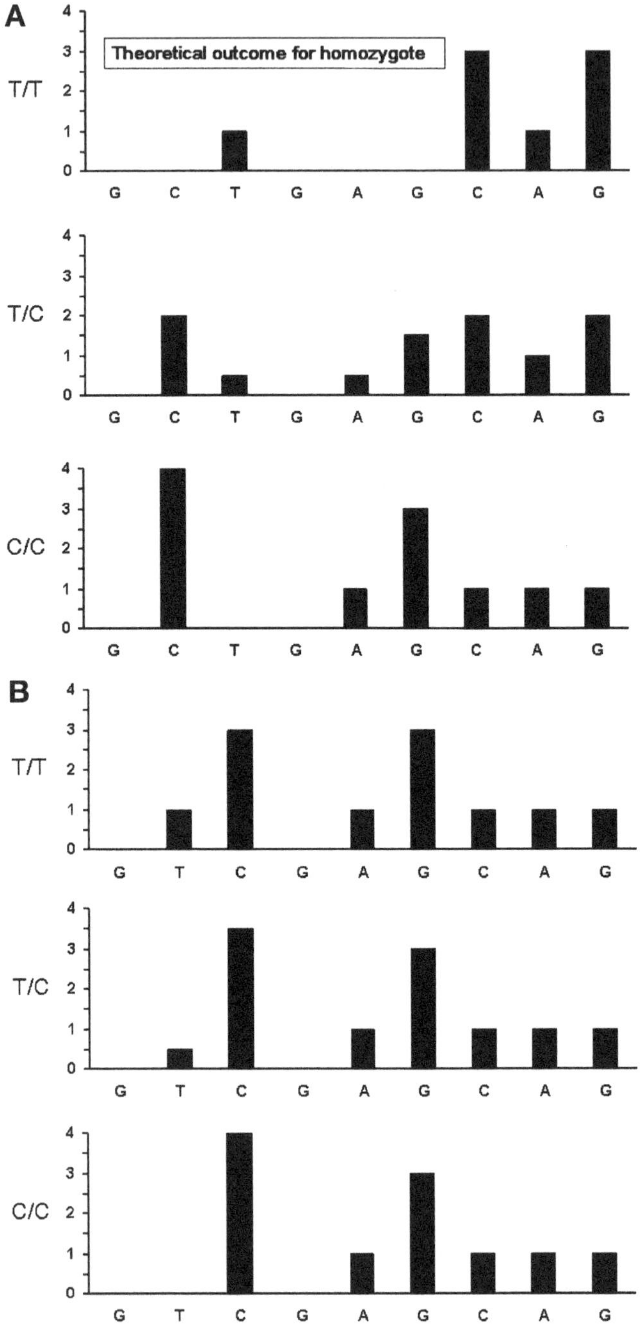

Fig. 2. Theoretical peak heights for marker rs3180018 in the *SCAMP 3* gene. **(A)** The dispensation order GCTGAGCAG; heterozygous individuals show changes in three peaks (C, T, and A). **(B)** The dispensation of the C and T is reversed (GTCGAGCAG), and only two peaks change (C and T). The sequence to analyze was T/CCCCAGGGCAGCATG.

3.3. RNA Extraction, cDNA Genotying, and Allele Quantification

Isolated tissues are snap frozen in liquid nitrogen and stored at –80°C until use. Cell cultures have been extracted directly using either Trizol reagent (Invitrogen Life Technologies, Carlsbad, CA) or Qiagen RNEasy kits (Qiagen, Valencia, CA). Adipose tissue was extracted using the Qiagen RNeasy Lipid Tissue Minikits. We have also successfully used muscle for the protocols described using the Biotecx Ultraspec RNA isolation system (Biotech, Pasadena, CA). Genomic DNA must be removed, particularly if heteronuclear RNA will be examined. Treatment of 50-μg aliquots with DNAse I (Ambion, Inc., Austin, TX) removes any contaminating genomic DNA. The concentration of purified total RNA was determined spectrophotometrically at 260 nm, and quality and quantity checked on an Agilent 2100 Bioanalyzer. Alternatively, the RNA 260 to 280 n*M* ratio should be greater than 1.7, and the quality should be checked on a 1.8% agarose gel.

First-strand cDNA synthesis was performed with the Taqman Reverse Transcription RT kit (Applied Biosystems, Inc., Foster City, CA) or Superscript II (Invitrogen Inc.). Although we have used random hexamers for reverse transcription, theoretically sequence variants might alter reverse transcription resulting in allelic bias that would not be observed in the genomic DNA. Validation with oligo-dT-initiated reverse transcription is thus recommended. For the Taqman Reverse Transcription Kit, the reaction mixture contained 5 m*M* $MgCl_2$, 1X reverse transcription buffer, 500 μ*M* dNTD mixture, 0.4 U/μL RNase inhibitor, 1.25 U/μL reverse transcriptase, 2.5 μ*M* of random hexamer primers, and 500 ng of total RNA and sterile H_2O to a final volume of 30 μL. The reaction was incubated at 25°C for 10 min, followed by reverse transcription at 48°C for 30 min and final heat inactivation of the enzyme at 95°C for 5 min. After cooling on ice for 5 min, the cDNA was stored at –20°C until use.

3.4. Allele Quantification

After reverse transcription, cDNA (30 μL) was diluted 1:20 in sterile water. We added 5 μL of the diluted cDNA to the total 25 μL PCR mixture using Master Mix (Promega Inc.). A universal sequence was appended to one primer of each set (5'-CACGACGTTGTAAAACGAC-3' gene-specific primer). PCR was done in the presence of 1.25 m*M* forward and reverse primer and 3.75 m*M* biotinylated universal primer. PCR conditions included an initial denaturation at 94°C for 3 min, then 60 cycles of 94°C for 30 s, 56°C for 45 s, and 72°C for 45 s, with a final extension of 6 min. All reactions are conducted in 96-well PCR plates.

After the PCR reaction, the product was mixed with 3 μL streptavidin-coated polystyrene beads (GE Healthcare Life Sciences Division, Piscataway, NJ) and 25 μL 2X binding buffer. The mixture was agitated at 1400 rpm on a

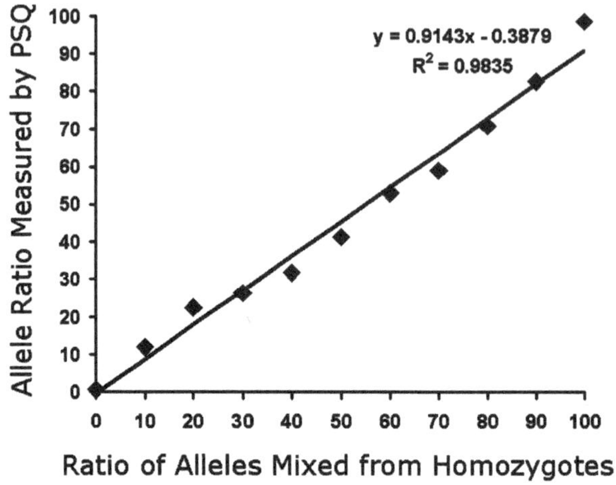

Fig. 3. After reverse transcription, cDNA concentrations were determined by picogreen for homozygous individuals. Calibration samples for the single-nucleotide polymorphism (SNP) rs3180018 in the *SCAMP3* gene were prepared by mixing homozygous DNA samples at different ratios (2:8, 3:7, 4:6, 5:5, 6:4, 7:3, 8:2). The graph shows the mixed ratio on the *x*-axis and the observed ratio on the *y*-axis. Although the relationship is linear ($r^2 = 0.98$), the slope deviates sufficiently from one that a 50:50 ratio actually measures 60:40. Pipetting errors are a clear limitation, and use of genomic DNA is greatly preferable.

Thermomixer (Brinkmann Instruments, Inc., Westbury, NY) for 10 min, after which the beads were lifted from the plate with the hand Pyrosequencing Vacuum Tool (Pyrosequencing, Inc) and rinsed for 5 s each in 70% ethanol, 0.8% NaOH, and, finally, 10 m*M* Tris (pH 7.6). Beads with bound DNA were released from the vacuum and suspended in 30 μL of annealing buffer with 3 μL of 3 m*M* sequencing primer in the 96-well PSQ plate. The plate was denatured for 2 min (80°C) and loaded on the PSQ-96 (Pyrosequencing) following the usual protocol. Genomic DNA was treated identically for genotyping.

Each allele of individuals heterozygous was quantified by Pyrosequencing using SNP Software AQ (Pyrosequencing, Inc.). Peak heights for each allele were compared with peak heights estimated for the same assay in samples of genomic DNA from heterozygous individuals. Alternatively, if the same assay could not be used on cDNA and genomic DNA, homozygous samples of cDNA were used to construct a mixing curve with ratios 0, 10, 20, 30, 40, and 50% of each allele (12 points; *see* **Fig. 3**). Each measure was conducted in duplicate.

3.5. Statistical Analysis

The allelic ratio in cDNA must be normalized to remove systematic artifacts caused by unequal amplification or biases in peak heights caused by inequalities of light emission from incorporation of different nucleotides. The simplest method is to compare directly to the DNA ratio in heterozygous individuals. We generally test at least 20 heterozygous individuals for the genomic DNA, in which the allele ratios are expected to be 50:50. However, some assays are impossible to design for both cDNA and genomic DNA, particularly if the SNP is near the intron:exon boundary. In this case, a standard curve may be constructed as illustrated in **Fig. 3** by mixing cDNA from individuals homozygous for each allele in ratios of 0/10, 1/9, 2/8, 3/7, and so on. The curve is expected to be linear, and thus a linear regression line is easily drawn and the "corrected" ratio calculated using the linear regression formula. This method does not, however, provide an easy test of statistical significance.

We have used two tests of statistical significance, but both have limitations in the absence of strong linkage disequilibrium. First, cDNA and genomic DNA ratios can be transformed to a normal distribution, usually using a natural logarithmic transformation. Significance can be tested by unpaired *t*-test when genomic DNA comes from different individuals (the usual case in our studies), or paired *t*-test if the tissue and genomic DNA is from the same samples. The latter has the advantage that insertions or deletions in genomic DNA will be identified. Alternatively, with small numbers of heterozygous individuals, the distribution may be impossible to determine. In this case, a nonparametric test performs much better, and we have used the Mann–Whitney U test in SPSS v12 for Intel Processors (SPSS, Inc., Chicago, IL). We have considered $p < 0.05$ to be significant. However, as previously noted, in many cases only a few heterozygotes will show distorted allelic ratios. In this setting, tests of statistical significance may not have meaning. An alternative proposed by Pastinen et al. ***(19)*** is to establish 95% confidence intervals for the allelic ratio based on genomic DNA from heterozygous individuals. Clearly, this approach requires access to a large number of heterozygous individuals, but if 20 subjects are indeed available, a distribution can be determined using standard statistical methods. An observed ratio that falls outside of the 95% confidence intervals is thus considered significant, and if three individuals fall outside, validation steps are initiated.

3.6. Selected Results

To illustrate the previous methods, we show results from our own studies of genes for which we have examined the allelic balance in both genomic DNA and cDNA from transformed lymphocytes and adipose tissue. The reader is referred to other published studies in which we have used the methods described previously for additional examples ***(20,27,28)***. Interleukin (IL)-6 levels and sequence variants have been implicated in type 2 diabetes and insulin resistance. IL-6 acts through the interleukin-6 receptor (IL-6R), which comprises two subunits, IL-6R and gp130. IL-6R confers the specificity to IL-6 action and maps to human chromosome 1q21, a region of well-replicated linkage to type 2 diabetes. We identified 11 variants with a minor allele frequency over 5%, and these included both coding and 3' untranslated variants in Caucasian and African-American individuals. To test whether sequence variation in *IL-6R* might alter the mRNA ratio between the alleles, we examined the allelic balance for four SNPs: D358A, V1385I, and two 3' untranslated region SNPs. As we previously described, we corrected for unequal amplification of the two alleles by comparing the same assay in genomic and cDNA, and we compared the ratios using a nonparametric, Mann–Whitney U test. Our results are illustrated in **Fig. 4** and presented in **Table 1**. Consistent with the arguments made previously, some but not all heterozygous individuals showed increased expression of one allele by 50–250%, although the mean levels were only modestly overexpressed at 30–60% excess expression from one allele.

As we previously note, the large variation in expression that we observed is consistent with the possibility that the SNPs we examined are in fact tagging a regulatory SNP at some distance. One means to test this hypothesis is to compare the number of individuals with extreme values of the ratio. Hence, we analyzed the data to determine the number of samples in genomic DNA and cDNA showing extreme values of at least a 50% increased level for one allele (corrected ratio 60/40%, 1.5). Although the mean ratios are only modestly increased, the number of extreme values for expression ratios was markedly increased for transformed lymphocyte mRNA for all four SNPs (p-values by χ^2 0.0003 to <0.0001; **Table 1**), and were also significant for adipocyte cDNA for the two 3' untranslated region SNPs (p <0.0007). One unexpected finding was that some DNA samples had reproducibly distorted ratios, and those ratios were reflected in the lymphocyte cDNA suggesting copy number variation.

In the second example, we examined SNP rs3180018 in the *SCAMP3* gene on chromosome 1q. In this case, as previously noted, we had to correct the ratio using a mixing curve (**Fig. 3**), and we chose a dispensation order that would provide only two peaks (**Fig. 2**). Using the regression line observed = 0.91(known ratio) + 0.39 from **Fig. 3**, we obtained allelic ratios of 61.6 ± 2.3 in adipose, 66.9 ± 3.8 in muscle, and 58.7 ± 7.9 in transformed lymphocytes. Although formal statistical tests are impossible in this setting because different assays had to be used in the known standard, the calibration curve should remove any systematic bias. The samples in this case consistently show the same ratio, which is in accordance with the extensive linkage disequilibrium in this region and the presence of only two common, extended haplotypes. Notably, the haplotypes in this case are associated with type 2 diabetes in several populations.

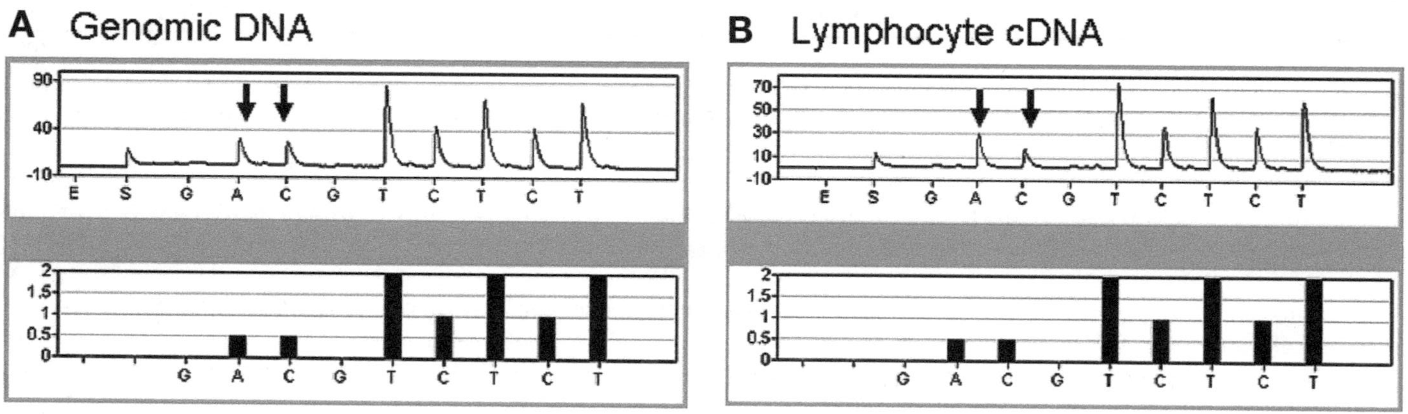

Fig. 4. PSQ allele quantification of genomic DNA **(A)** and cDNA **(B)** for single-nucleotide polymorphism (SNP) rs8192284 (D358A) in exon 9 of the *IL-6R* gene. The ratio in genomic DNA shows the theoretical 50:50 ratio (actual measured 50.2 and 49.8). In contrast, the cDNA shows an altered ratio of 63.8% G and 36.2% A, or a ratio of 1.76 in transformed lymphocytes. Because the genomic DNA is at the theoretical ratio, no correction is required.

Table 1
Allele-Specific mRNA Levels in Transformed Lymphocytes and Adipose mRNA

SNP ID	Ratio, genomic DNA (%)	Ratio, lymphocyte cDNA (%)	Ratio, lymphocyte cDNA (corrected)	DNA samples with ratio > 1.5 (%)	Lymphocyte samples with ratio > 1.5 (%)	*p*-value lymphocyte cDNA vs genomic DNA	Ratio, adipocyte cDNA	Adipocyte samples with ratio > 1.5 (%)	*p*-value adipocyte cDNA vs genomic DNA
D358A (A/C; rs8192284)	52.78/47.21	62.53/37.47	1.71 (0.70)	4/267 (1.5%)	11/19 (57.9%)	<0.000001	1.06 (0.23)	4/32 (12.5%)	0.00004
SNP12 G/A (V385I, rs2228146)	49.65/50.36	42.92/57.07	1.45 (0.65)	0/97 (0%)	3/10 (30%)	0.006	NA	NA	NA
SNP13 A/G (3UTR, rs2229238)	45.88/54.12	39.49/60.52	1.46 (0.57)	1/122 (0.8%)	7/18 (38.9%)	<0.000001	1.13 (0.40)	5/23 (21.7%)	0.010
SNP16 T/C (3UTR, rs4072391)	54.79/45.20	57.92/42.07	1.29 (0.53)	3/234 (1.3%)	7/20 (35.0%)	0.001	1.099 (0.63)	6/26 (23.1%)	0.011

Imbalance in allele-specific expression for four SNPs in the transcribed sequence, using the same assay for RNA and DNA. Significance of lymphocyte allelic ratio when compared with DNA was determined by Mann–Whitney U test.

Acknowledgments

The preparation of this chapter and the work cited from our laboratory was supported by the Research Service of the Department of Veterans Affairs, by National Institutes of Health (NIH) grants DK39311 and DK54636, and by the a grant to the University of Arkansas for Medical Sciences General Clinical Research Center M01 RR14288 from the National Centers for Research Resources (NIH).

References

1. Cheung, V. G., Spielman, R. S., Ewens, K. G., et al. (2005) Mapping determinants of human gene expression by regional and genome-wide association. *Nature* **437,** 1365–1369.
2. Monks, S. A., Leonardson, A., Zhu, H., et al. (2004) Genetic inheritance of gene expression in human cell lines. *Am. J. Hum. Genet.* **75,** 1094–1105.
3. Farh, K. K., Grimson, A., Jan, C., et al. (2005) The widespread impact of mammalian MicroRNAs on mRNA repression and evolution. *Science* **310,** 1817–1821.
4. Bento J. L., Palmer, N. D., Mychaleckyj, J. C., et al. (2004) Association of protein tyrosine phosphatase 1B gene polymorphisms with type 2 diabetes. *Diabetes* **53,** 3007–3012.
5. Grant, S. F., Thorleifsson, G., Reynisdottir, I., et al. (2006) Variant of transcription factor 7-like 2 (TCF7L2) gene confers risk of type 2 diabetes. *Nat. Genet.* **38,** 320–323.
6. Horikawa, Y., Oda, N., Cox, N. J., et al. (2000) Genetic variation in the gene encoding calpain-10 is associated with type 2 diabetes mellitus. *Nat. Genet.* **26,** 163–175.
7. Mohlke, K. L. and Boehnke, M. (2005) The role of HNF4A variants in the risk of type 2 diabetes. *Curr. Diab. Rep.* **5,** 149–156.
8. Ben Asher, E. and Lancet, D. (2004) 5-Lipoxygenase activating protein (ALOX5AP): association with cardiovascular infarction and stroke. *Isr. Med. Assoc. J.* **6,** 318–319.
9. Helgadottir, A., Manolescu, A., Thorleifsson, G., et al. (2004) The gene encoding 5-lipoxygenase activating protein confers risk of myocardial infarction and stroke. *Nat. Genet.* **36,** 233–239.
10. Helgadottir, A., Manolescu, A., Thorleifsson, G., et al. (2004) The gene encoding 5-lipoxygenase activating protein confers risk of myocardial infarction and stroke. *Nat. Genet.* **36,** 233–239.
11. Sayers, I., Barton, S., Rorke, S., et al. (2003) Promoter polymorphism in the 5-lipoxygenase (ALOX5) and 5-lipoxygenase-activating protein (ALOX5AP) genes and asthma susceptibility in a Caucasian population. *Clin. Exp. Allergy* **33,** 1103–1110.
12. Van Eerdewegh, P., Little, R. D., Dupuis, J., et al. (2002) Association of the ADAM33 gene with asthma and bronchial hyperresponsiveness. *Nature* **418,** 426–430.

13. Vermeire, S. and Rutgeerts, P. (2005) Current status of genetics research in inflammatory bowel disease. *Genes Immun.* **6,** 637–645.
14. Baier, L. J., Permana, P. A., Yang, X., et al. (2000) A calpain-10 gene polymorphism is associated with reduced muscle mRNA levels and insulin resistance. *J. Clin. Invest.* **106,** R69–R73.
15. Yan, H., Yuan, W., Velculescu, V. E., et al. (2002) Allelic variation in human gene expression. *Science* **297,** 1143.
16. Morley, M., Molony, C. M., Weber, T. M., et al. (2004) Genetic analysis of genome-wide variation in human gene expression. *Nature* **430,** 743–747.
17. Stranger, B. E., Forrest, M. S., Clark, A. G., et al. (2005) Genome-wide associations of gene expression variation in humans. *PLoS. Genet.* **1,** e78.
18. Buckland, P. R. (2005) The importance and identification of regulatory polymorphisms and their mechanisms of action. *Biochim. Biophys. Acta.* **1762,** 17–28.
19. Pastinen, T., Sladek, R., Gurd, S., et al. (2004) A survey of genetic and epigenetic variation affecting human gene expression. *Physiol. Genomics* **16,** 184–193.
20. Wang, H., Zhang, H., Jia, Y., et al. (2004) Adiponectin receptor 1 gene (ADIPOR1) as a candidate for type 2 diabetes and insulin resistance. *Diabetes* **53,** 2132–2136.
21. Wang, H., Zhang, Z., Chu, W., et al. (2005) Molecular screening and association analyses of the interleukin 6 receptor gene variants with type 2 diabetes, diabetic nephropathy, and insulin sensitivity. *J. Clin. Endocrinol. Metab.* **90,** 1123–1129.
22. Bray, N. J., Buckland, P. R., Owen, M. J., et al. (2003) Cis-acting variation in the expression of a high proportion of genes in human brain. *Hum. Genet.* **113,** 149–153.
23. Lo, H. S., Wang, Z., Hu, Y., et al. (2003) Allelic variation in gene expression is common in the human genome. *Genome Res.* **13,** 1855–1862.
24. Ge, B., Gurd, S., Gaudin, T., et al. (2005) Survey of allelic expression using EST mining. *Genome Res.* **15,** 1584–1591.
25. Knight, J. C., Keating, B. J., Rockett, K. A., et al. (2003) In vivo characterization of regulatory polymorphisms by allele-specific quantification of RNA polymerase loading. *Nat. Genet.* **33,** 469–475.
26. Das, S. K., Chu, W., Zhang, Z., et al. (2004) Calsquestrin 1 (CASQ1) gene polymorphisms under chromosome 1q21 linkage peak are associated with type 2 diabetes in Northern European Caucasians. *Diabetes* **53,** 3300–3306.
27. Elbein, S. C., Zheng, H., Jia, Y., et al. (2004) Molecular screening of the human glutamine-fructose-6-phosphate amidotransferase 1 (GFPT1) gene and association studies with diabetes and diabetic nephropathy. *Mol. Genet. Metab.* **82,** 321–328.
28. Zhang, H., Jia, Y., Cooper, J. J., et al. (2004) Common variants in glutamine:fructose-6-phosphate amidotransferase 2 (GFPT2) gene are associated with type 2 diabetes, diabetic nephropathy, and increased GFPT2 mRNA levels. *J. Clin. Endocrinol. Metab.* **89,** 748–755.
29. Sun, A., Ge, J., Siffert, W., et al. (2005) Quantification of allele-specific G-protein beta3 subunit mRNA transcripts in different human cells and tissues by Pyrosequencing. *Eur. J. Hum. Genet.* **13,** 361–369.

30. Gruber, J. D., Colligan, P. B., and Wolford, J. K. (2002) Estimation of single nucleotide polymorphism allele frequency in DNA pools by using Pyrosequencing. *Hum. Genet.* **110,** 395–401.
31. Wasson, J., Skolnick, G., Love-Gregory, L., et al. (2002) Assessing allele frequencies of single nucleotide polymorphisms in DNA pools by Pyrosequencing technology. *BioTechniques* **32,** 1144–1150.
32. Pastinen, T., Ge, B., Gurd S., et al. (2005) Mapping common regulatory variants to human haplotypes. *Hum. Mol. Genet.* **14,** 3963–3971.
33. Gabriel, S. B., Schaffner, S. F., Nguyen, H., et al. (2002) The structure of haplotype blocks in the human genome. *Science* **296,** 2225–2229.
34. Gibbs, R. A., Belmont, J. W., Hardenbol, P., et al. (2003) The International HapMap Project. *Nature* **426,** 789–796.
35. Xie, X., Lu, J., Kulbokas, E. J., et al. (2005) Systematic discovery of regulatory motifs in human promoters and 3' UTRs by comparison of several mammals. *Nature* **434,** 338–345.

14

Pyrosequencing® Genotype Storage Techniques

Derek J. Van Booven

Summary

Data storage and data coordination are important aspects of project design and execution. Pyrosequencing® technology allows thousands of data-points to be collected per day. Consequently, a consistent and reliable method of data input and storage is vital. This chapter discusses the strengths and weaknesses of data storage systems.

Key Words: Data storage; spreadsheet; relational database; genotype.

1. Introduction

So you have genotypes. What do you do now? It is a good idea to think about how you want to store the data before you start data collecting. This step is too often overlooked, and depending on the amount of data available, can lead to an extraordinary amount of time spent manually manipulating the data to a manageable format.

Data overload is of primary concern with data storage. Raw data is not very useful if it is in a storage system that cannot be filtered and looked at it in a logical manner. The data becomes useful information when it is managed properly. Questions asked about data can be solved much easier with a functional system of queries made to the data source.

2. Data Storage Techniques

2.1. Spreadsheets

The best way to have the data the way you want it is to format it yourself. Spreadsheets are primarily used because of their ease of use and familiarity (everyone has used a spreadsheet at one point). Spreadsheets help out in basic data storing models, but in the case of genotype information they fall short of their actual usefulness. Also, with spreadsheets there are many existing con-

From: *Methods in Molecular Biology, vol. 373: Pyrosequencing® Protocols*
Edited by: S. Marsh © Humana Press Inc., Totowa, NJ

cerns or problems that even having the best formatting genius in the world working at your side does not make up for.

2.2. Problems With Spreadsheets

There are four major problems with spreadsheets. The first of which is spreadsheet usage. Popular spreadsheet programs have file locks on them that only allow one person at a time to use them. The spreadsheet is restricted and the files are only able to be read and not modified. An independent copy of the same database is therefore the best solution and then merging the changes back into the main spreadsheet is tedious and can get very messy. Some newer versions have shared files, but this adds another level of complexity for your spreadsheet.

The second major problem with spreadsheets is data integrity. Functions that manipulate the data (such as basic sorting) do not always work, and can lead to headaches from trying to get back to the original starting point. Also, manually inputting data into a spreadsheet can lead to errors of various kinds. It is very easy to click on the wrong cell and paste in the data, but if the cell was not the right one then the data could be off for individual samples. Also, manual naming of the variants or the assay can lead to the wrong data being put in the wrong place within the spreadsheet. Manual entry should be avoided at all costs.

The third major problem with spreadsheets is data validation. Input masks are not always easy to create within a spreadsheet. Depending on the complexity of the spreadsheet, there will always be validation issues that are difficult to handle within a single spreadsheet. From this data formatting is also a key issue. For example, the difference between Apr 21st, 21 Apr, April 21, 4/21, 21-4-2006 do not look like major issues, but down the road can lead to headaches inputting data.

Finally, a concern with spreadsheets is data duplication. If there are any cases of redundant data in the spreadsheet, then the data should not be in a spreadsheet. The issue here is not one of disk space (as the cost of disk space is relatively cheap), but one of tracking. For example, when storing sample information, it is not a good idea to have this stored in a spreadsheet as sample information (such as freezer location) can change and thus have to be changed in every instance of the spreadsheet.

A relational database is an excellent solution to these problems. By using a relational database many different kinds of rules and checks can be applied in an easier way rather than dealing with a spreadsheet. For example, if a primer is inputted into the database, it is very easy to attach a script on the database that will look up the individual gene sequence and confirm the location of the primer and the location of the variant. This can be confirmed rather than relying on manual entry into the spreadsheet of a primer sequence.

3. Relational Databases

A relational database is defined as a database that maintains a set of separate related files (tables), but combines data elements from the files for queries and reports when queried. These queries that drive the database are those defined by statements in a structured query language. Through structured query language statements the database is asked a question to which the answer comes much easier than looking it up in a table manually or more efficient than searching for it in a flat file.

Relational databases have two major advantages over flat spreadsheets. First, they eliminate redundancy. If a relational database is properly designed then there should be no place where data is duplicated. Second, relational databases greatly improve the time taken to find data. The speed at which a relational database can find information far exceeds the time taken to go through an entire database looking for that one needle in a haystack. Third, database management is easier to handle through a relational database rather than a spreadsheet. Because maintenance averages at about 70% of a program's lifespan, the relational database approach improves efficiency long term. The three core database principles of integrity, uniqueness, and efficiency are achieved better through a relational database rather than a spreadsheet.

4. Database Design

4.1. Definitions

Relational databases are primarily made up of tables. A table is an entity that can best be described as a spreadsheet. Similar to a spreadsheet, it has rows (called records), and columns (called fields). The power behind these tables is to relate the table in a way to other tables that creates a level of complexity, but allows values to be used as a lookup to additional information. A record is one in which there is information associated with it. A field is the information associated with that record. For example, a person will have a name associated with them. The person is the record, but the name is the field in which their name is stored.

4.2. Introduction to Data Modeling

As the name implies, "relational"databases have relationships. Between each table there exists a common element that feeds off each other that we call keys. A primary key (PK) is a key in which only one element exists within that table. Most primary keys are unique identifiers that could be a single numeric value or a complex set of characters that are made of multiple sources of information. For a field to be a PK it must have a non-null value, the value must be unique for each record, and the values must not change or become null during

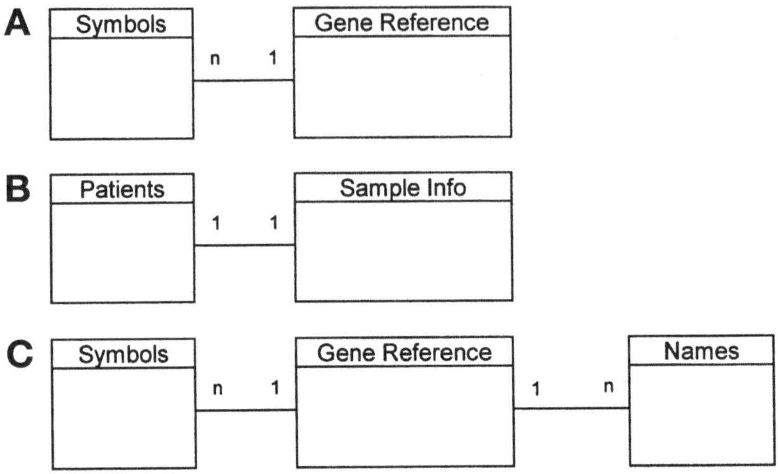

Fig. 1. Relationship types.

the existence of the record. In our example, a PK could be the gene reference field in the main gene table. There will only be one unique value for every gene record in the main gene table. A foreign key (FK) is a key that completes the uniqueness of a particular record within a table. An FK in our example can be a value that declares the gene type. There is a gene reference that is the PK, and the gene type that is the FK. Every dependent field in a table must be foreign keys.

4.3. Relationship Types

There are three primary types of relationships in relational database design (**Fig. 1**). It is important to understand the types of relationships before even looking at designing a database.

4.3.1. One to One

The rarest type of relationship is 1:1 (**Fig. 1A**). This is rare because logically if you have one record associated with exactly one record in a different table then perhaps you should have put that field in the other table. However, this is not always the case. If we accept the fact that multiple samples cannot come from the same person, a 1:1 relationship could be a single sample record associated with a single patient record. Patient information is recommended to be protected at high security so the patient information is associated with a sample record only and should be kept in a separate table.

4.3.2. One to Many

Perhaps the most common form of relationship is a one-to-many (1:N) relationship (**Fig. 1B**). This occurs when a single record can be associated with many records. For example, there are potentially many alleles (two in most single-nucleotide polymorphism [SNP] cases, but potentially more in cases such as insertion/deletions [INDELs] or variable nucleotide tandem repeats [VNTRs]) per a single SNP. Another example is there are many variants per a single gene. Considering the overloading issues listed previously, it is imperative to pay more attention to this type of relationship more than anything else. Most central tables from which the entire relational database is created will have at least a single 1:N relationship.

4.3.3. Many to Many

A many-to-many (N:N) relationship occurs when there are many associated records with many associated records in another table (**Fig. 1C**). This also appears commonly throughout our model. For example, there are possibly many unique Unigene identifiers for a single gene while there are many common gene names. For example, *ABCB1* has a current Unigene number of Hs.489033, but in the past the same gene has been classified under the Unigene number of Hs.21330. There are also many common names for *ABCB1*. Some include *p*-glycoprotein and ATP-binding cassette subfamily B member one. The proper symbol has even changed from *MDR1* to *ABCB1*. In this example there are many identifiers for a single gene, but this is not easily linked within a normal data model. In order to define a N:N relationship there must be an intermediate table between the tables in which the data fields are associated. As defined in **Fig. 1C**, you can see the gene name table is associated with the Unigene identifiers through the gene reference table. This is the proper way to define the N:N relationship.

5. An Example Entity Relationship Model for Genotype Data

5.1. Overall Description

A database can be as simple or as complex as its creator wants it to be. It can be as simple as a 2-table spreadsheet with a secondary lookup table, or it can be a 20-table intermediate database, or it can be a monstrous database that requires maintenance. An example of an entity relationship model for storing genotypes is presented in **Fig. 2**. There are five distinct areas of data that is stored in this database. There is gene data, variant data, assay data, genotype data, and sample data.

The structure of the database is fairly simple. There is gene data that is linked to variant data. The variant data table branches off into assay description and genotype data. Genotype data then branches off into sample information and plate setup. This is a fairly common setup for genotyping data that has been tested and works well with Pyrosequencing data.

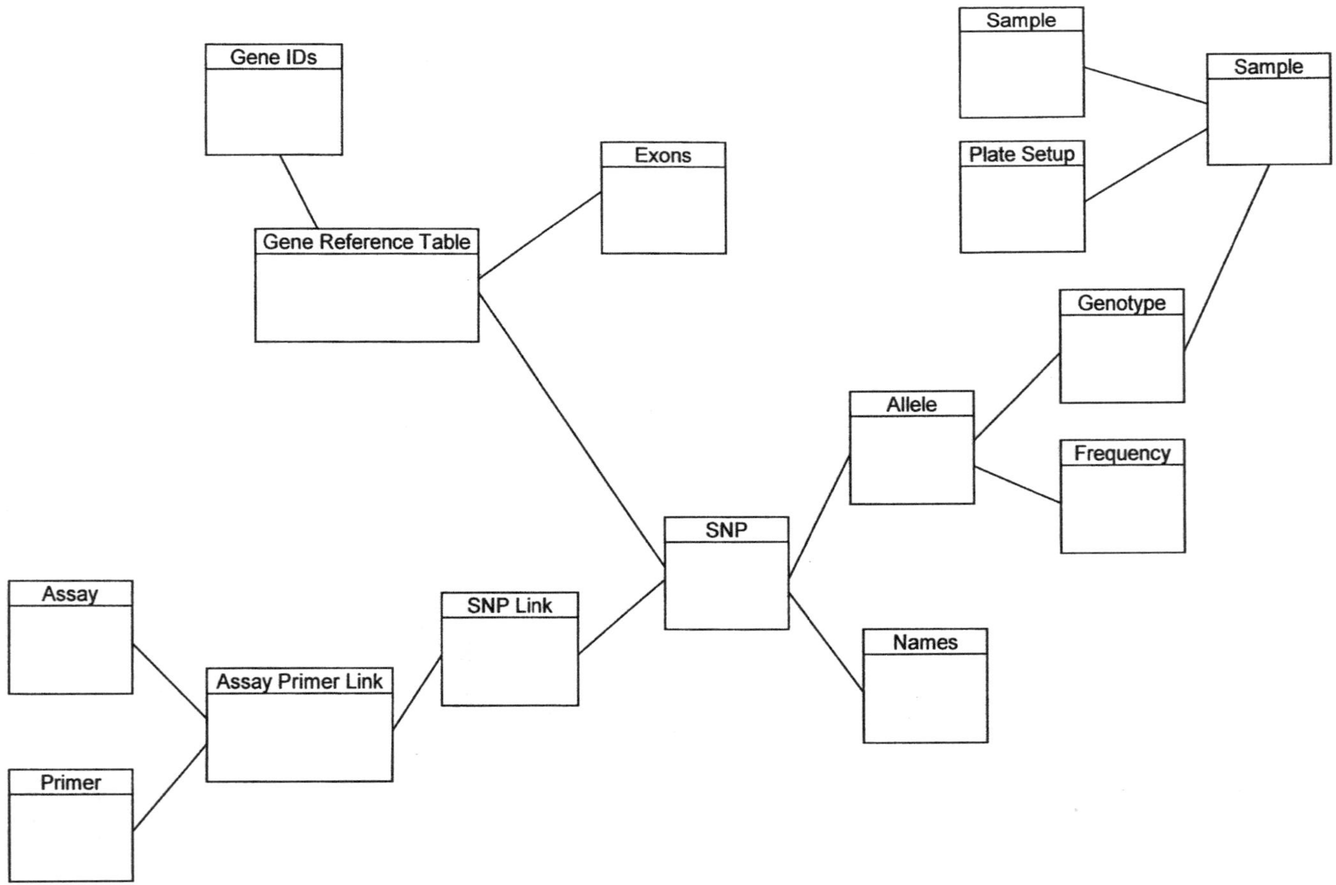

Fig. 2. Entity relationship model for genotype information.

5.2. Gene Information

There are a host of data sources publicly available. Depending on individual resources, there are possibly many related identifiers in different databases. For example, there could be multiple gene symbols or multiple common gene names for a particular gene. The advantage to having a single gene table and the other gene identifier tables on the outside comes back to the N:N relationship definition. It is necessary to have that core table linking many types of identifiers to a single gene. Each value can be distinguished as a primary value with the addition of a single "yes/no" field within the table.

Singular data about a gene can go within the gene reference table. For example, chromosomal location would be stored within the gene reference table as the location of the gene is not likely to change. Most information associated with a specific gene will be variable depending on the sequence used. Even the number of exons can be variable within the same gene because of alternative splicing. Thus, most of the information about genes will be kept outside the gene reference table.

5.3. Variant Information

This is perhaps the most central table of the entire model. Everything runs off this table and is linked through this table. In other words, you cannot get from a sample to a gene without going through this table. This is a core table that has extreme importance. There are identifiers related to the variant in the form of the names table that branches off it. Possible identifiers from external databases, alternative names from literature, proper names, and local names of these variants force a separate table. Again, as with the gene values, a simple "yes/no" field can distinguish which one is a primary name to use.

Static data can be stored within the main variant data table. For example, a field that stores the verification status if the variant has been validated can be stored within this main table as well as other data for certain variant types, such as a codon number if the amino acid changes within the variant sequence. Information such as that should be stored in the central variant. Other information, such as alleles and genotype information, are stored outside this main table, but linked through a common identifier.

5.4. Genotype Information

The most useful information that is stored within a genotype database is genotypes. Genotype information is useful for either individual genotypes or from allele or genotype frequencies that are derived from a specific sample set. This database schema handles both cases. There is a frequency table that is derived from the allele table in which frequencies from outside sources (external databases or literature) can be stored. There is also a genotype table in which individual genotypes are stored from individual genotyping experiments.

The genotype table is the main table in which the genotypes are stored from the Pyrosequencing system. An intermediary parser is required to take the genotypes that are exported from the Pyrosequencing system into a form that can be stored within our data storage model. From here the data can be more easily manipulated and updated to answer all genotyping questions. The genotype information table consists of two links; one to the variant allele table to the genotype table and one to the sample information table. From the Pyrosequencing system there will be two records, which will be stored in the genotype table signifying the two chromosomes making up an individual genotype.

However, the previously mentioned information shows bad data management practice. We are duplicating records in the genotype table for each genotype that turns out to be homozygous. A solution to this problem, while still going through the allele table, is to have the individual expected genotypes stored within the allele table. This would eliminate the duplication within the genotype table, but still duplicates information stored within the allele table. There is no point to having individual alleles when the genotype is the value we really want to explore.

Alternatively, a third solution to this problem would be for the genotype table to link directly to the SNP table and to bypass the allele table altogether. This might be the soundest strategy, but in doing this you will lose the dependency on the allele table and for genotypes to be manually inputted risking the danger of manual error. In other words, the double-checking step that comes through the allele table is lost and dependency on input is high. This might be an acceptable risk in most cases though.

5.5. Sample Information

Off the genotype table is a link to the sample information part of the database. Typically there will be a single genotype for a single sample. Often there will be instances in which the genotype could not be determined. For cases like this, the database should not store such information, but rather can be deduced from queries within the database in which those genotypes are missing from samples known to have been run. Thus, the need for control over the Pyrosequencing plate setup and individual sample sets.

Pyrosequencing plate setup is kept separate in their own table primarily owing to multiple samples being placed on the different plate setups. This table is very basic consisting of only a static sample ID and a well position indicating where on the Pyrosequencing plate the sample comes from. Sample sets can be linked to this table; however, it is better for a sample set to be defined through the sample information table rather than the Pyrosequencing plate setup table.

A particular problem with this schema in general is when individual samples are genotyped and will not necessarily have a definable Pyrosequencing plate setup. These cases happen when so-called "gap" plates consisting of indeterminate genotypes or sample sets consisting of less than a full plate are placed onto a single plate. Many plate setups are created from this and do not necessarily have to be stored within the database. However, if the plate setups are required to be stored within the database then another level should be put in here with a Pyrosequencing plate name table and a primary field in that table for those in which we can easily query the database for the current Pyrosequencing plate setups.

5.6. Assay Information

It seems counterintuitive to have the assay information so far away from the Pyrosequencing plate setup table, but in our model it works to have it this way. The individual assay is set up with three primers. The possibility exists of not only a N:N relationship between the assay and primer, but also another N:N relationship between the assay/primer relationship and the variant. In essence, this is a N:N:N relationship. The possibility exists for a single variant to be on multiple assays and for multiple primers to be associated with both multiple assays and multiple variants. In other words, everything can be associated multiply.

Also contained within this general assay is specific primer information associated with the assay. There will be at least three primers per assay. If information about a certain primer is found to be faulty then the primer record can be replaced. Multiple primers can be associated with multiple assays, as well as multiple variants. It is important to first establish assay and primer matchups before we link this to the variants. This is because generally the assays and the primers are contained in tables with a closer relationship. The variant can be added later, but the important part of defining an assay are the primers associated with the assay.

6. Database Software Design Issues

6.1. Security Concerns

It is commonly accepted that there is genetic variation between populations. Although every population will have different allele or genotype frequencies at one or more loci, there is also considerable variation among members of the same population. There is considerable danger when using this information.

Individual genotypes can lead to discrimination, for example, in determining which people get certain treatments. Insurance companies can see that an individual might have a predisposition to a certain disease and refuse to insure the individual as a result of their genotype. Population genotypes can also be discriminating. Rival tribes or warring nations that have the same frequency of genotypes might come under the notion that they want to be nothing like their counterparts and thus genotyping is wrong.

The bottom line is there needs to be considerable security in the genotyping database setup. Individual patients have to be blinded as much as possible, and the overall genotype frequencies have to be delicately handled. Security policies about how to handle such data is also highly recommended for all IT personnel and violations of such policies have to be dealt with. It is a primary responsibility for IT personnel to have complete control of the data, and to provide the proper security measures necessary for data protection and integrity.

7. Conclusion

Spreadsheets are okay, but relational databases are better. Databases are not hard, and if properly thought about to start with can save a huge amount of time and efforts long term. Questions can be asked as simply as obtaining an overall genotype frequency of a specific population set, or correlating a genotype with a specific phenotype or clinical outcome. With a good database design, analysis programs can be developed to analyze the data or have a measure of quality control with the genotypes measured.

If the relational database is set up properly, then these questions can be answered immediately and effectively.

Index

Q,R

S

FSC
www.fsc.org
MIX
Papier | Fördert
gute Waldnutzung
FSC® C083411

Kolibri 360 GmbH
Ferdinand-Jühlke-Straße 7
99095 Erfurt, Deutschland
produktsicherheit@kolibri360.de